Further physiology

8

Genes and biotechnology

9

More about microbes

10

Use, damage and repair

11

Animal behaviour

This book and your GCSE course

	AQA	EDEXCEL 360
Web address	www.aqa.org.uk	www.edexcel.org.uk
Specification number	4411	2105
Modular tests	Biology 1 – 45 min structured question test. Or Biology 1a and 1b – 2 x 30 minute objective tests 25% Biology 2 – 45 min short answer test 25% Biology 3 – 45 min short answer test 25%	2 x 20 min objective tests covering B1a and B1b 20%. One objective test and one short answer test covering B2 20% Short answer test covering B3 30%
Terminal papers	none	none
Availability of exams	Nov, March, June for Biology 1a Jan and June for Biology 2 and 3	Nov, March, June (B3 only in June).
Coursework	25%	30%
BIOLOGY		
Our bodies in action	B1.11.1	B1b3
Health and disease	B1.11.2 B1.11.3 B1.11.4	B1b4
Variation and genetics	B1.11.6 B1.11.7	B1a1 B1a2
Organisms and the environment	B1.11.5 B1.11.8	B1a1
Cells and growth	B2.11.6, B2.11.8, B2.11.2, B2.11.3	B2.3, B2.1, B2.2
Ecology	B2.11.4, B2.11.3, B2.11.5	B2.3
Further physiology	B2.11.7, B2.11.6	
Genes and biotechnology	B2.11.8, B2.11.6	B2.1, B2.2
More about microbes	B3.13.5, B3.13.4, B3.13.6	B3.1
Use, damage and repair	B3.13.1, 3.13.3, B3.13.4	
Animal behaviour		B3.2

Visit your awarding body for full details of your course or download your complete GCSE specifications.

Use these pages to get to know your course

- *Make sure you know your exam board*
- *Check which specification you are doing*
- *Know how your course is assessed:*
  - *what format are the papers?*
  - *how is coursework assessed?*
  - *how many papers?*

OCR A	OCR B
www.ocr.org.uk	
J633	J643
2 x 40 min objective style tests. One paper tests B1,B2,B3 and the other B4,B5,B6. 33.3%	2 x 60 min structured tests. One paper tests B1,B2,B3 and the other B4,B5,B6. 66.7%
Ideas in context including B7 33.3%	none
Jan, June	Jan, June
33.3%	33.3%
B3.3	B1a, B1b, B1d, B1f
B2.1, B2.2, B2.3, B2.4	B1a, B1b, B1c, B1e
B1.1, B1.2, B1.3, B1.4, B3.1	B1g, B1h, B2f
B3.4	B1b, B1c, B2a, B2b, B2c, B2d, B2e, B2g, B2h
B4.2, B5.2, B5.1, B5.3	B3a, B4a, B3d, B3e, B3h, B3b
	B4e, B4f, B4d, B4g, B4h
B4.1, B4.4, B6.1, B6.2, B6.3, B6.4, B6.5, B6.6	B3c
	B3a B3g
B7.4	B6a, B6b, B6c, B6d, B6h
B7.7, B7.6, B7.5, B7.3	B5a, B5c, B5d, B5e, B5f, B5h

Preparing for the examination

Planning your study

The last few months before taking your final GCSE examinations are very important in achieving your best grade. However, the success can be assisted by an organised approach throughout the course. This is particularly important now as all the science courses are available in units.

- After completing a topic in school or college, go through the topic again in your Revise GCSE Biology Study Guide. Copy out the main points on a sheet of paper or use a highlighter pen to emphasise them.
- Much of memory is visual. Make sure your notes are laid out attractively using spaces and symbols. If they are easy to read and attractive to the eye, they will be easier to remember.
- A couple of days later, try to write out these key points from memory. Check differences between what you wrote originally and what you wrote later.
- If you have written your notes on a piece of paper, keep this for revision later.
- Try some questions in the book and check your answers.
- Decide whether you have fully mastered the topic and write down any weaknesses you think you have.

Preparing a revision programme

Before an external examination, look at the list of topics in your examination board's specification. Go through and identify which topics you feel you need to concentrate on. It is a temptation at this time to spend valuable revision time on the things you already know and can do. It makes you feel good but does not move you forward.

When you feel you have mastered all the topics, spend time trying sample questions that can be found on your examination board's website. Each time, check your answers with the answers given. In the final week, go back to your summary sheets (or highlighting in the book).

How this book will help you

Revise GCSE Biology Study Guide will help you because:

- it contains the essential content for your GCSE course without the extra material that will not be examined
- it contains GCSE Exam Practice Questions to help you to confirm your understanding
- examination questions from 2007 are different from those in the past. Trying past questions will not help you when answering some parts of the questions in 2007. The questions in this book have been written by experienced examiners.
- the summary table will give you a quick reference to the requirements for your examination

Four ways to improve your grade

1. Read the question carefully

Many students fail to answer the actual question set. Perhaps they misread the question or answer a similar question they have seen before. Read the question once right through and then again more slowly. Some students underline or highlight key words in the question as they read it through. Questions at GCSE contain a lot of information. You should be concerned if you are not using the information in your answer.

2. Give enough detail

If a part of a question is worth three marks you should make at least three separate points. Be careful that you do not make the same point three times. Draw diagrams with a ruler and label with straight lines.

3. Correct use of scientific language

There is important scientific vocabulary you should use. Try to use the correct scientific terms in your answers. The way scientific language is used is often a difference between successful and unsuccessful students. As you revise, make a list of scientific terms you meet and check that you understand the meaning of these words. Learn all the definitions. These are easy marks and they reward effort and good preparation.

4. Show your working

All science papers include calculations. Learn a set method for solving a calculation and use that method. You should always show your working in full. Then, if you make an arithmetical mistake, you may still receive marks for correct science. Check that your answer is given to the correct number of significant figures and give the correct units.

How Science Works

From 2007, all GCSE science courses must cover certain factual detail, similar to the detail that has been required for many years. Now, however, each course must also include study of '**How Science Works**'.

This includes four main areas:

- **Data, evidence, theories and explanations**
  This involves learning about how scientists work, the differences between data and theories and how scientists form theories.

- **Practical skills**
  How to test a scientific idea including collecting the data and deciding how reliable and valid it is.

- **Communication skills**
  Learn how to present information in graphs and tables and to be able to analyse information that has been provided in different forms.

- **Applications and implications of science**
  Learning about how new scientific discoveries become accepted and some of the benefits, drawbacks and risks of new developments.

The different examining bodies have included material about how science works in different parts of their examinations. Often it is in the coursework but you are also likely to come across some questions in your written papers. Do not panic about this and think that you have not learnt this work. Remember these questions test your skills and not your memory; that is why the situations are likely to be unfamiliar. The examiners want you to show what you know, understand and can do.

To help you with this, there are sections at the end of each chapter called **How Science Works** and questions about how science works in the **Exam Practice Questions**. This should give you an idea of what to expect.

Chapter 1

Our bodies in action

The following topics are covered in this chapter:

- Food and digestion
- Respiration
- Response to stimuli
- Homeostasis
- Hormones and reproduction

1.1 Food and digestion

What is in a balanced diet?

OCR B B1b

All organisms require **food** to survive. It provides energy and the raw materials for growth. We take our food in ready-made as complicated organic molecules. These food molecules can be placed into seven main groups.

A **balanced diet** needs the correct amounts of each of these types of food molecules.

Remember a balanced diet is the correct amount of each food, not simply 'enough'.

Food type	Made up of	Use in the body
carbohydrates	starch and sugars, e.g. glucose	supply or store of energy
fats	fatty acids and glycerol	rich store of energy
proteins	long chains of amino acids	growth and repair
minerals	different elements, e.g. iron	iron is used to make haemoglobin
vitamins	different structures, e.g. vitamin C	Vitamin C prevents scurvy
fibre	cellulose	prevents constipation
water	water	all chemical reactions take place in water

The exact amount of each substance that is needed in a balanced diet will vary. It depends on how **old** the person is, whether they are **male or female** and how **active** they are.

For example, teenagers need a high-protein diet to provide the raw materials for growth. You can estimate the recommended daily average (RDA) protein intake for a person using the formula:

$$\text{RDA in g} = 0.75 \times \text{body mass in kg}$$

There are differences between the sexes because of the time of the growth spurt and due to periods in girls.

However, it is not only the amount of protein that is important but also the type. Proteins from animals are called **first class proteins** because they contain more variety of **amino acids** compared with plant proteins.

Some people's diet may be influenced by other factors than just their daily requirements. Some people may be vegetarians or vegans and some religions require certain diets to be followed. Some people may have to avoid certain foods to prevent them becoming ill.

Digestion and absorption

OCR B B1b

KEY POINT **The job of the digestive system is to break down large food molecules. This is called digestion.**

Digestion happens in two main ways: **physical** and **chemical**. Physical digestion occurs in the mouth where the teeth break up the food into smaller pieces.

Remember enzymes are biological catalysts found in all cells of the body.

Chemical digestion is caused by digestive enzymes that are released at various points along the digestive system.

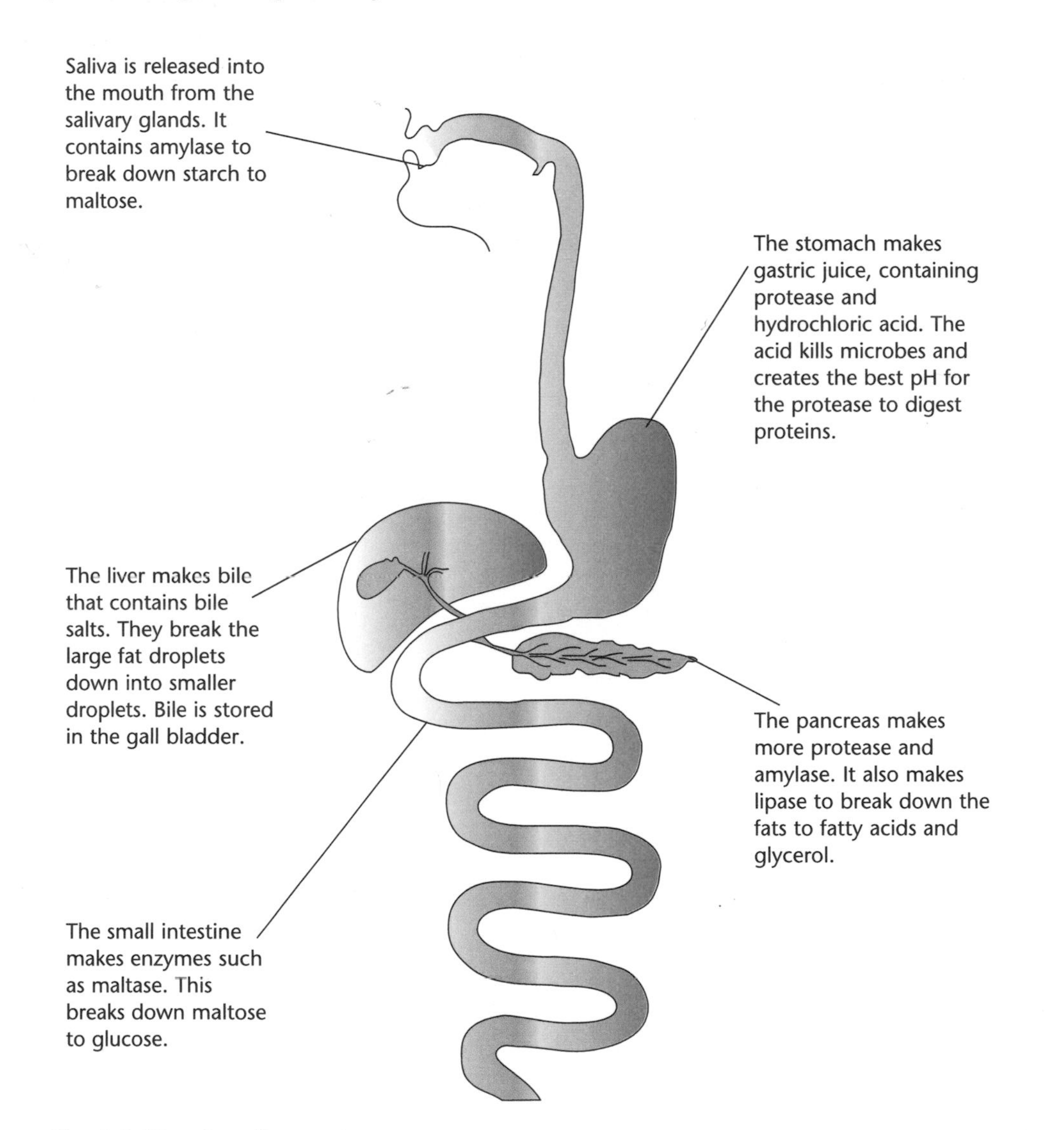

Fig. 1.1 The digestive system.

Once the food molecules have been digested, they are small enough to diffuse into the bloodstream or lymph vessels. This is called **absorption**.

1.2 Respiration

Aerobic and anaerobic respiration

OCR B B1b

Aerobic respiration

Aerobic respiration is when glucose reacts with oxygen to release energy. Carbon dioxide and water are released as waste products.

glucose + oxygen → carbon dioxide + water + **energy**

$$C_6H_{12}O_6 + 6O_2 \rightarrow 6CO_2 + 6H_2O + \text{energy}$$

We use the energy released from respiration for many processes. Respiration also gives off heat, which is used to maintain our high body temperature. Our rate of respiration can be estimated by measuring how much oxygen we use.

During exercise, the body needs more energy and so the rate of respiration increases. The breathing rate increases to obtain extra oxygen and remove carbon dioxide from the body. The heart beats faster so that the blood can transport the oxygen and carbon dioxide faster. This is why our pulse rate increases.

It is actually the build up of carbon dioxide that makes us breathe faster.

Anaerobic respiration

KEY POINT

When not enough oxygen is available, glucose can be broken down by anaerobic respiration. This may happen during hard exercise.

In humans:

glucose → lactic acid + **energy**

Being able to respire without oxygen sounds a great idea. However, there are two problems:

- Anaerobic respiration releases less than half the energy of that released by aerobic respiration.
- Anaerobic respiration produces lactic acid. Lactic acid causes muscle fatigue and pain.

The build up of lactic acid is called the **oxygen debt**. After the exercise is finished, extra oxygen is needed by the liver to remove the lactic acid.

1.3 Response to stimuli

Patterns of response

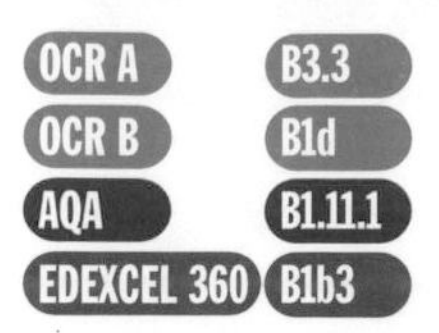

Plants can also respond to stimuli but the response is usually slower than that of animals.

All living organisms need to respond to changes in the environment.

Although this happens in different ways the pattern of events is always the same:

stimulus → detection → co-ordination → response

Detecting the stimulus

KEY POINT **Receptors are specialised cells that detect a stimulus. Their job is to convert the stimulus into electrical signals in nerve cells.**

Some receptors can detect several different stimuli but they are usually specialised to detect one type of stimulus:

Stimulus	Type of receptor
light	photoreceptors in the eye
sound	vibration receptors in the ears
touch, pressure, pain and temperature	different receptors in the skin
taste and smell	chemical receptors in the tongue and nose
position of the body	receptors in the ears

A **sense organ** is a group of receptors gathered together with some other structures.

The other structures help the receptors to work more efficiently. An example of this is the eye.

Co-ordination

The body receives information from many different receptors at the same time.

KEY POINT **Co-ordination involves processing all the information from receptors so that the body can produce a response that will benefit the whole organism.**

In most animals this job is done by the central nervous system (CNS).

Response

KEY POINT **Effectors are organs in the body that bring about a response to the stimulus.**

Usually these effectors are muscles and they respond by contracting. They could however be glands and they may respond by releasing an enzyme. Many responses are **reflexes**.

An example of a sense organ – the eye

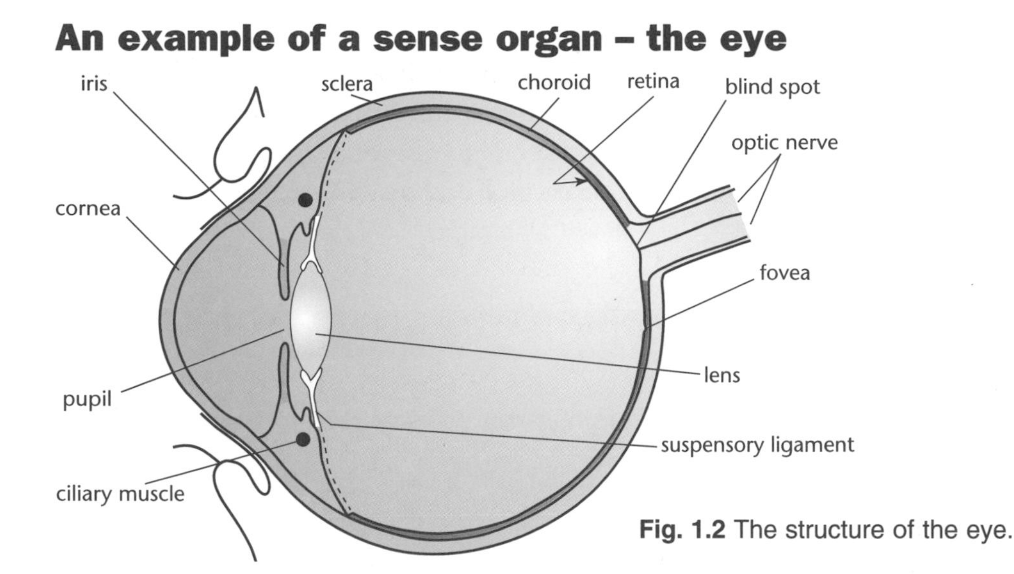

Fig. 1.2 The structure of the eye.

The light enters the eye through the pupil. It is focused onto the **retina** by the **cornea** and the **lens**. The size of the **pupil** can be changed by the muscles of the **iris** when the brightness of the light changes. The aim is to make sure that the same amount of light enters the eye. The job of the lens is to change shape so that the image is always focused on the light-sensitive retina.

Too much bright light entering the eye could damage the retina.

The receptors are cells in the retina called rods and cones. They detect light and send messages to the brain along the **optic nerve**.

The lens must be a different shape when the eye looks at a close object compared with a distant object. This is to make sure that the light is always focused on the back of the retina. The ciliary muscle changes the shape of the lens as shown in the diagram. This is called **accommodation**.

Remember, when the muscles are relaxed the eye is focused on a distant object.

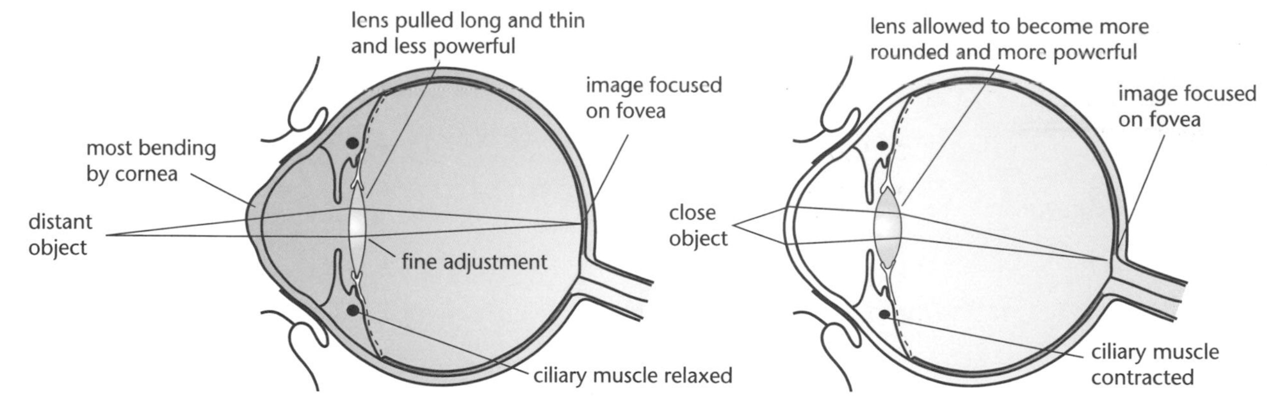

Fig. 1.3 How the eye focuses.

Some people have problems with their eyes:

Condition	Cause	Treatment
long or short sight	the eyeball or lens is the wrong shape	long sight and short sight can be corrected by wearing convex or concave lenses respectively; cornea surgery can now also be used
red–green colour blindness	lack of certain cones in the retina	no treatment
poor accommodation	lens becomes less elastic in senior citizens	wearing glasses with half convex and half concave lenses

The eyes are also used to judge **distances**. Animals that hunt usually have their eyes on the front of their head. Each eye has a slightly different image of the object. This is called binocular vision and it can be used to judge distance. Animals that are hunted usually have eyes on the side of their heads. This gives monocular vision and they cannot judge distances so well. They can, however, see almost all around.

Neurones and synapses

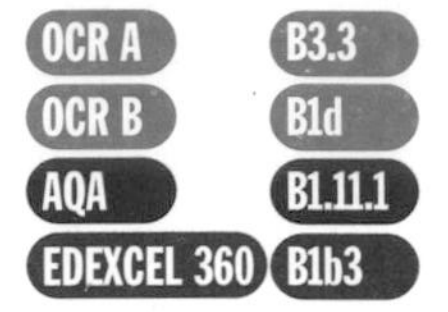

To communicate between receptors and effectors the body uses two main methods.

These are:

- **nerves**
- **hormones**.

A neurone is a specialised cell that is adapted to pass electrical impulses.

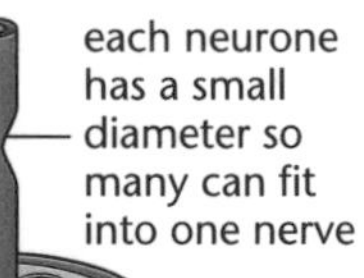

Fig. 1.4 The structure of a nerve.

The **central nervous system** (CNS) contains millions of neurones but outside the CNS, neurones are grouped together into bundles of hundreds or thousands. These bundles are called nerves.

There are different types of neurones.
The three main types of neurones are:

- **sensory neurones** – they carry impulses from the receptors to the CNS

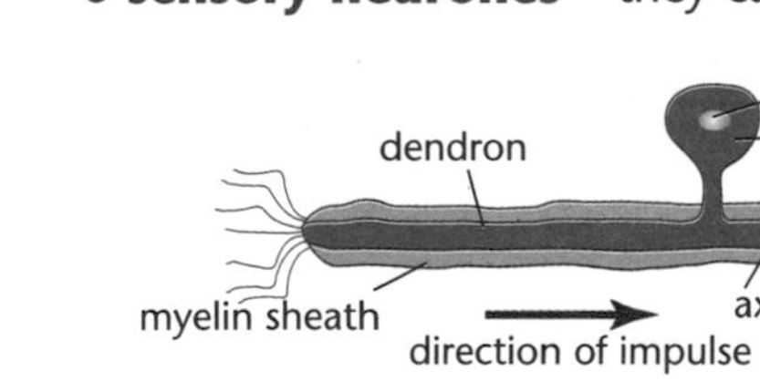
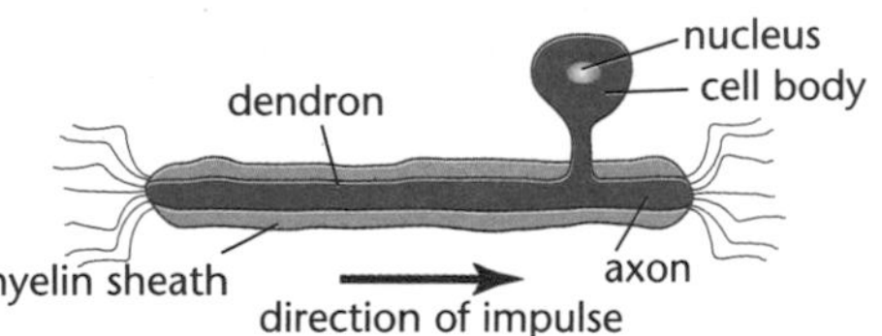

Fig. 1.5 Sensory neurone.

- **motor neurones** – they carry impulses from the CNS to the effectors

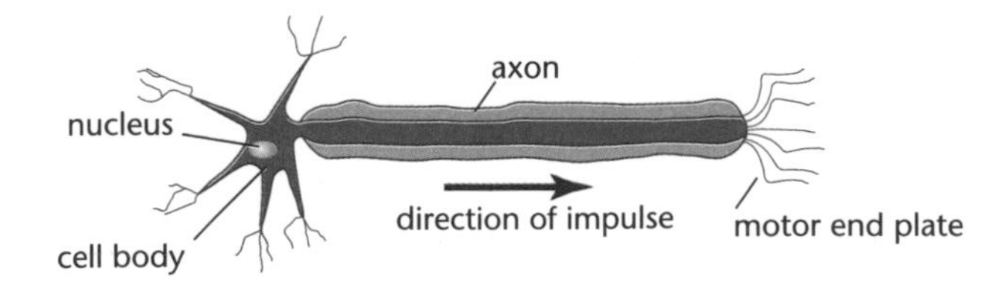

Fig. 1.6 Motor neurone.

- **relay neurones** – they pass messages between neurones in the CNS.

Exam questions often ask how neurones are adapted for their job.

Although all neurones have different shapes, they all have certain features in common:

- One or more long **projections** from the cell body to carry the impulse a long distance.
- A fatty covering (**myelin sheath**) around the projection for insulation.
- Many fine endings (**dendrites**) so that the impulse can be passed on to many cells.

Synapses

Each neurone does not directly end on another neurone.

There is a small gap between the two neurones called a **synapse**.

In order for an impulse to be generated in the next neurone, a **chemical transmitter** is released. This then diffuses across the small gap.

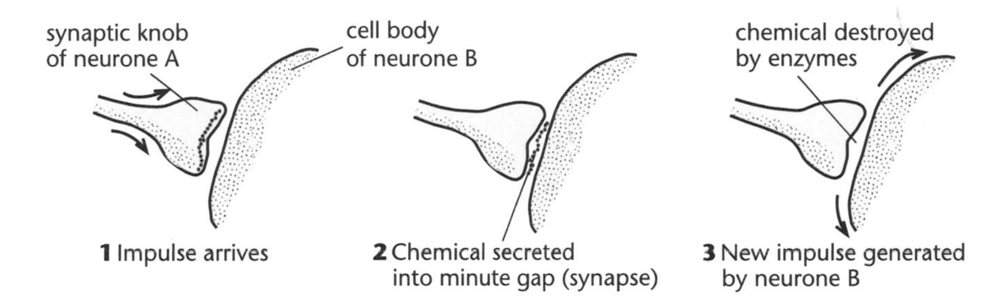

Fig. 1.7 Chemical transmission between nerves.

Many drugs work by interfering with synapses. They may block or copy the action of neurotransmitters in certain neurones.

Types of response

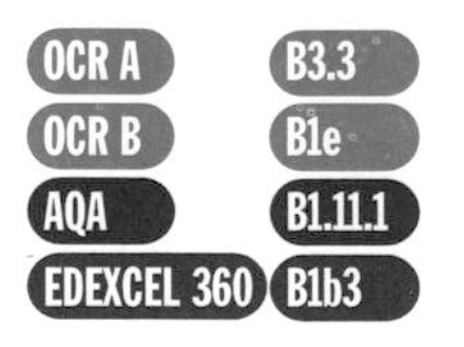

The nervous system is made up of the CNS and the **peripheral nervous system**.

KEY POINT

The CNS is the brain and spinal cord. The peripheral nervous system is all the nerves passing information to and from the CNS.

Once the information reaches the CNS from a sensory neurone there is a choice:
Either: The message may be passed straight to a motor neurone via a relay neurone. This is very quick and is called a **reflex action**.

Or: The message can be sent to the higher centres of the brain and the organism might decide to make a response. This is called a **voluntary action**.

A reflex action

Do not say that reflexes do not involve the brain. Reflexes such as the pupil reflex go via the brain.

All reflexes are:

- fast
- do not need conscious thought
- protect the body.

Examples of reflexes include the knee jerk, pupil reflex, accommodation, ducking and withdrawing the hand from a hot object.
This diagram shows the pathway for a reflex that involves the spinal cord:

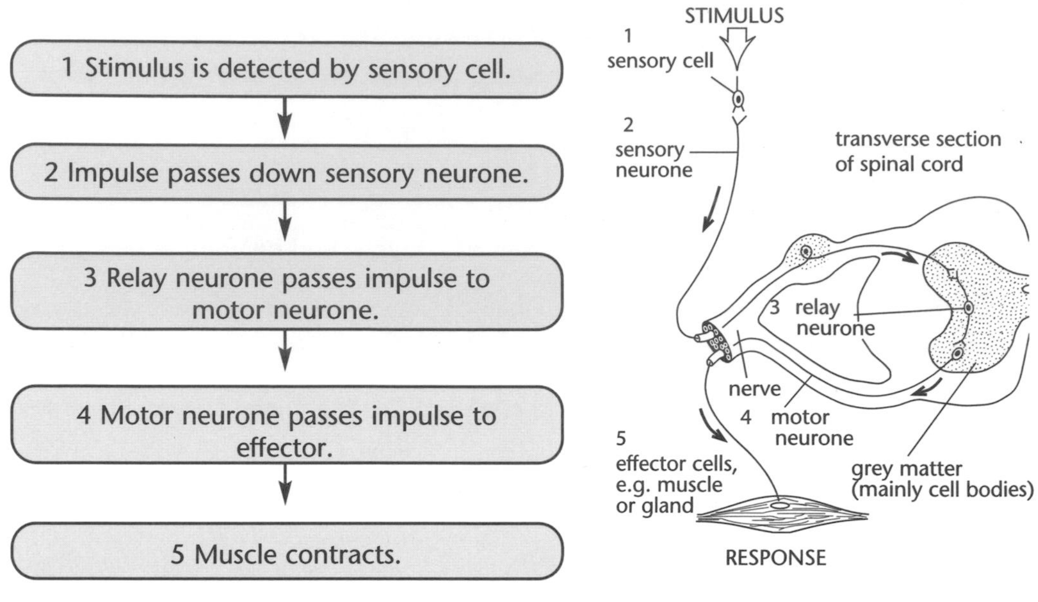

Fig. 1.8 A reflex action.

A voluntary action

Voluntary actions need a conscious decision in order to take place. They therefore always involve the **brain**.

Only Edexcel candidates need to know the structure of the brain.

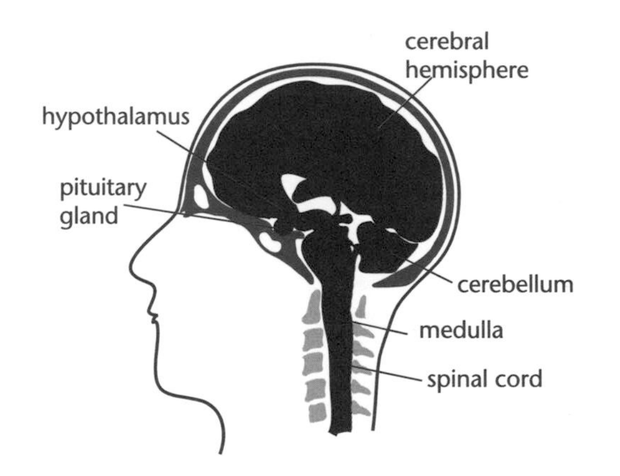

Fig. 1.9 The structure of the brain.

The **cerebral hemisphere** is the area of the brain where the decisions are made. Nerve impulses from here are sent down the spinal cord to effectors via motor neurones.

1.4 Homeostasis

Homeostasis and hormones

OCR A	B3.3
OCR B	B1f
AQA	B1.11.1
EDEXCEL 360	B1b3

KEY POINT **It is vital that the internal environment of the body is kept fairly constant. This is called homeostasis.**

The different factors that need to be kept constant include:

Many of the mechanisms that are used for homeostasis involve hormones.

KEY POINT **Hormones are chemical messengers that are carried in the blood stream.**

Hormones are carried dissolved in the plasma of the blood.

They are released by glands and pass to their target organ.

Figure 1.10 shows the main hormone-producing glands of the body. Between them they make a number of different hormones:

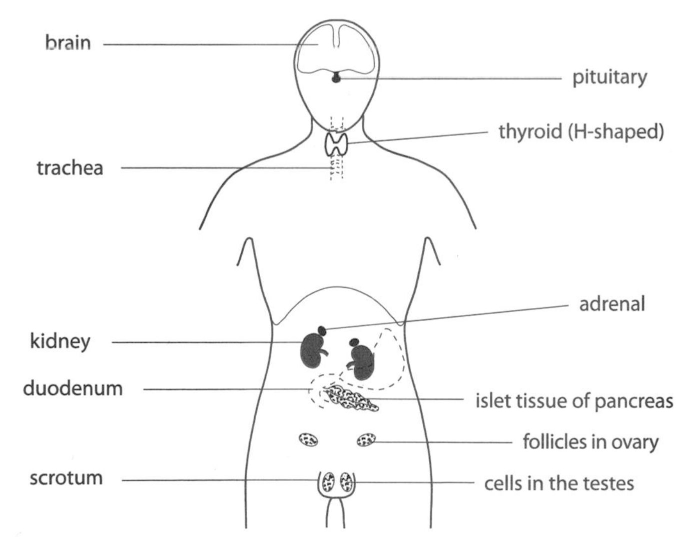

Fig. 1.10 Hormone-producing glands.

Many devices in the home work by negative feedback such as the thermostat that controls the temperature of an oven.

Hormones take longer to have an effect than nerves but their responses usually last longer.

Many of these control mechanisms work by **negative feedback**. This means that if the levels change too much, a hormone is released and this brings the change back to the normal level.

Control of blood sugar

It is vital that the sugar or glucose level of the blood is kept constant. If it gets too low then cells will not have enough to use for respiration. If it is too high then glucose may start to pass out in the urine.

Insulin is the hormone that controls the level of glucose in the blood.

If sugar passes out in the urine it takes water with it, resulting in large volumes of urine.

When glucose levels are too high, more insulin is made. The insulin converts excess glucose into glycogen to be stored in the liver:

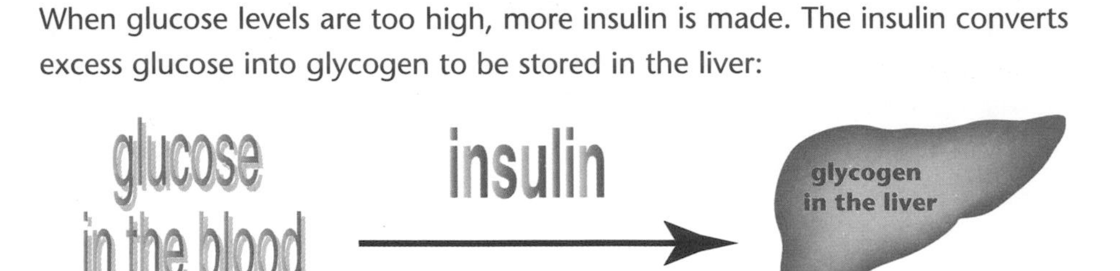

People with **diabetes** do not produce enough insulin naturally. They need regular insulin injections in order to control the level of glucose in their blood. They also need to control their diet carefully.

Control of body temperature

OCR B B1f

It is important to keep our body temperature at about 37 °C. This is because it is the best temperature for enzymes to work.

The blood temperature is monitored by the brain and if it varies from 37 °C, various changes are brought about.

Camels can put up with their body temperature rising by 6°C without their enzymes being damaged.

When we feel too hot

When we feel hot we need to lose heat faster, as our core body temperature is in danger of rising.

We do this by:

- **sweating** – as water evaporates from our skin, it absorbs heat energy. This cools the skin and the body loses heat.
- **vasodilation** – blood capillaries near the skin surface get wider to allow more blood to flow near the surface. Because the blood is warmer than the air, it cools down and the body loses more heat.

If the blood temperature gets too high it could lead to heat stroke and dehydration.

When we feel too cold

When we feel too cold we are in danger of losing heat too quickly and cooling down. This means we need to conserve our heat to maintain a constant 37 °C.

We do this by:

- **shivering** – rapid contraction and relaxation of body muscles. This increases the rate of respiration and more energy is released as heat
- **vasoconstriction** – blood capillaries near the skin surface get narrower and this process reduces blood flow to the surface. The blood is diverted to deeper within the body to conserve heat.
- **sweating less**.

If vasoconstriction occurs for a long time it can lead to frostbite.

Hypothermia occurs when the blood temperature gets too low. It can be fatal.

1.5 Hormones and reproduction

The reproductive hormones

OCR B B1f
EDEXCEL 360 B1b3

Hormones are responsible for controlling many parts of the reproduction process.

This includes:

- the development of the sex organs
- the production of sex cells
- controlling pregnancy and birth.

The main hormones controlling these processes are shown in the table.

Hormone	Male or female	Produced by the	Main function
testosterone	male	testes	stimulates the male secondary sexual characteristics
oestrogen	female	ovaries	stimulates the female secondary sexual characteristics; repair of the wall of the uterus; controls ovulation
progesterone	female	ovaries and placenta	prevents the wall of the uterus breaking down.

Testosterone and oestrogen control the changes occurring in the male and female bodies at puberty. These changes are the **secondary sexual characteristics**:

oestrogen
breasts grow,
pubic hair grows,
wide hips
develop

testosterone
body hair grows,
voice breaks,
muscle growth
increases

The secondary sexual characteristics also include the production of the sex cells. In the male they are **sperm** and in females they are **eggs**.

After puberty, sperm are produced continuously but in the female one egg is usually released about once a month.

This means that oestrogen and progesterone levels vary at different times in the monthly cycle.

Oestrogen levels are high in the first half of the cycle. The oestrogen prepares the wall of the uterus to receive a fertilised egg. It does this by making the wall thicker and increasing its supply of blood. It also triggers the release of an egg. This is called **ovulation**.

Progesterone is high in the second half of the cycle. It further repairs the wall of the uterus and stops it breaking down.

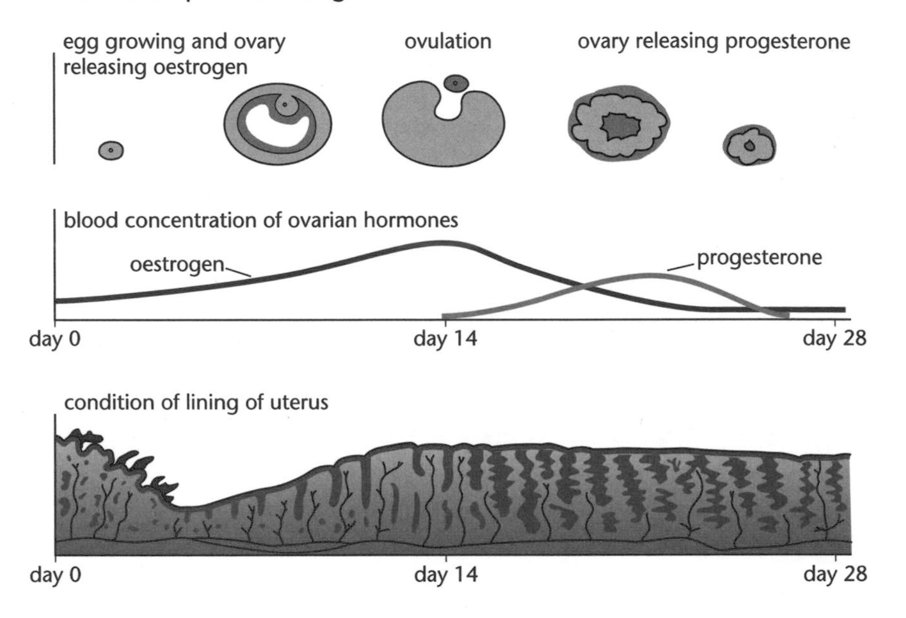

Fig. 1.11 Changes occurring during the monthly cycle.

Using hormones to control reproduction

The production of oestrogen and progesterone is controlled by the release of other hormones. These hormones are made in the **pituitary gland** in the brain.

OCR A and Edexcel candidates need to know how hormones can control reproduction but do not need to know the roles of FSH and LH.

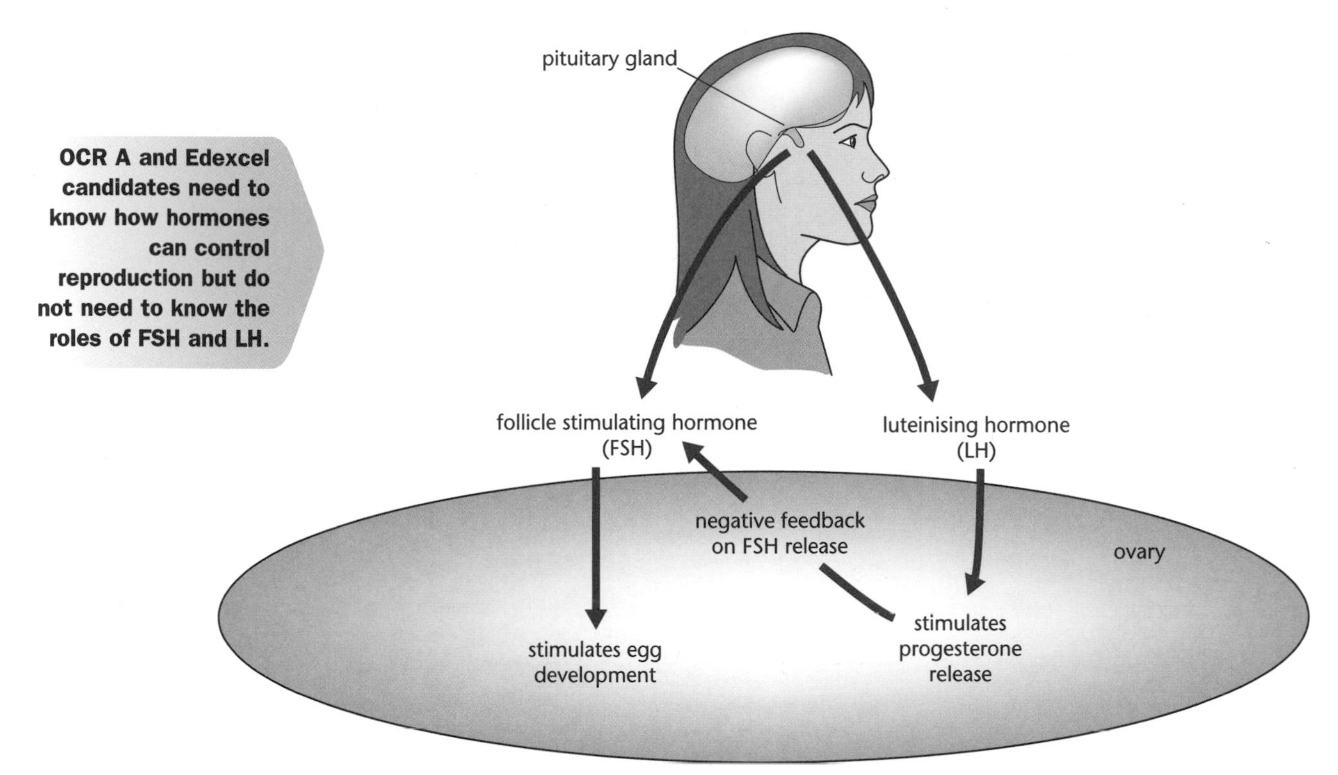

Fig. 1.12 Control of reproduction.

It is now possible to produce synthetic versions of these hormones. They can be used to control the fertility of women.

This can happen in two main ways.

Increasing fertility

Some women find it difficult to get pregnant because they do not produce eggs regularly. These women can take a **fertility drug**. This contains hormones that are similar to FSH. The drugs stimulate the production of eggs and sometimes a number of eggs are released each month.

Sometimes women are treated with fertility drugs and the eggs are removed from their body. The eggs can then be fertilised by sperm outside the body. The embryo can then be put back inside the uterus. This process is called *in vitro* fertilisation or IVF. It can be used on women who have blocked oviducts.

Decreasing fertility

Scientists are developing a male contraceptive pill that would stop sperm production.

Other women may want to stop themselves becoming pregnant. They take drugs that are called **oral contraceptives**. These drugs contain hormones that prevent the pituitary gland releasing FSH. This means that the ovary will not produce eggs.

HOW SCIENCE WORKS

OCR A	B2.4
OCR B	B1d, e
AQA	B1.11.3
EDEXCEL	B1b3, 4

Parkinson's disease and smoking – is it worth the risk?

Parkinson's disease is a disorder of the central nervous system that is caused by a loss of cells in an area of the brain.

Those cells produce dopamine, a chemical messenger responsible for transmitting signals across synapses. Loss of dopamine causes some nerve cells to fire out of control, leaving patients unable to control their movement in a normal way.

Patients often start off with slight shaking of the hands and they may eventually have difficulty walking, talking or completing other simple tasks.

The disease can be treated by giving a drug called L-dopa. The brain uses this chemical to make more dopamine. Unfortunately, the drug has a number of side effects and larger doses of the drug are often needed as the disease gets worse. Patients therefore have a difficult decision to make.

In recent studies on Parkinson's disease it has been suggested that there is a **correlation** between smoking and a reduced risk of having the disease.

Doctors are trying to find out if this means that smoking actually **causes** the protection.

Do these results mean that doctors are actually advising patients to smoke? No, they are not.

As a doctor said:

"The dangers of cigarette smoking far outweigh any as yet inconclusive evidence that there are advantages of protection from Parkinson's disease.

You need to look at the risks – smoking is the largest single cause of preventable death in many countries.

About 1 in 1000 people are likely to get Parkinson's disease but every year over 100 000 people die through cigarette smoking in the UK.

Smoking is just not worth the risk."

HOW SCIENCE WORKS

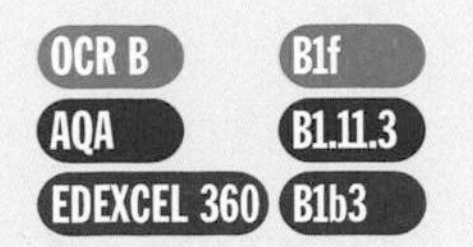

Parkinson's disease and smoking – is it worth the risk?

Parkinson's disease is a disorder of the central nervous system that is caused by a loss of cells in an area of the brain.

Those cells produce dopamine, a chemical messenger responsible for transmitting signals across synapses. Loss of dopamine causes some nerve cells to fire out of control, leaving patients unable to control their movement in a normal way.

Patients often start off with slight shaking of the hands and they may eventually have difficulty walking, talking, or completing other simple tasks.

The disease can be treated by giving a drug called L-dopa. The brain uses this chemical to make more dopamine. Unfortunately the drug has a number of side effects and larger doses of the drug are often needed as the disease gets worse. Patients therefore have a difficult decision to make.

Recently evidence has appeared that seems to show that people who smoke are less likely to have Parkinson's disease.

These studies show that there is a **correlation** between smoking and a reduced risk of having the disease. Doctor's are trying to find out if this means that smoking actually **causes** the protection.

Do these results mean that doctors are actually advising patients to smoke? No they are not.
As a doctor said:

"I think that fertility treatment by drugs or IVF treatment is too expensive and should not be paid for by the National Health Service."

"I think that people who are infertile have a right to have treatment. Being unable to have children can make them depressed and this is as serious as other illnesses."

HOW SCIENCE WORKS Questions

Treating fertility can produce some difficult ethical issues.

1. Explain the difficult choices facing a woman who has become pregnant with a large number of embryos. [2]
2. Suggest why some people think that it is not a good idea for women aged over 60 to have children. [2]

Exam practice questions

1. Which hormone in the list causes a boy's voice to deepen at puberty?
   A insulin
   B oestrogen
   C progesterone
   D testosterone **[1]**

2. Which of the following are female secondary sexual characteristics?
   A growth of breasts
   B sperm production
   C presence of ovaries
   D presence of a uterus **[1]**

3. Which of the following changes might happen in our body when we get too hot?
   A sweating increases
   B blood flow to the skin drops
   C we shiver more
   D blood vessels in the skin close up **[1]**

4. Finish the sentences using words from the list.

   absorption amino acids digested egested growth iron respired storing energy sugars vitamin C

   A balanced diet contains seven groups of substances. Proteins are necessary for ___1___ and are made from molecules called ___2___. Foods such as minerals are another group of substances. An example is ___3___, which is needed to make haemoglobin. Many molecules such as proteins are too large to be able to pass into our blood stream and so need to be ___4___ first. The process of taking food into the blood stream is called ___5___. **[5]**

5. The diagram shows some of the organs inside the body.

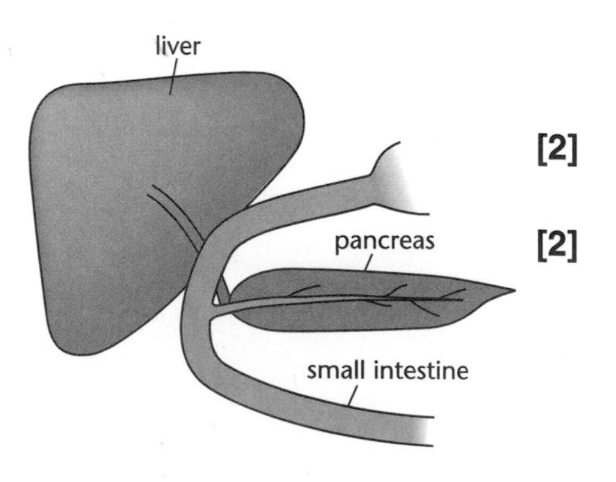

   (a) Write down the name of the organ shown that:
   (i) makes insulin
   (ii) stores glycogen **[2]**
   (b) It is very important to keep the level of glucose in the blood roughly constant. Explain why. **[2]**
   (c) A person has a drink of a liquid containing glucose.
   Explain the role of the organs shown in the diagram in dealing with this glucose.
   You should include:
   - how the glucose gets into the body
   - what happens when the blood glucose level rises
   - what happens to the glucose. **[5]**

Exam practice questions

6. The graph shows the level of two hormones in a woman and the woman's body temperature at different days of her monthly cycle.

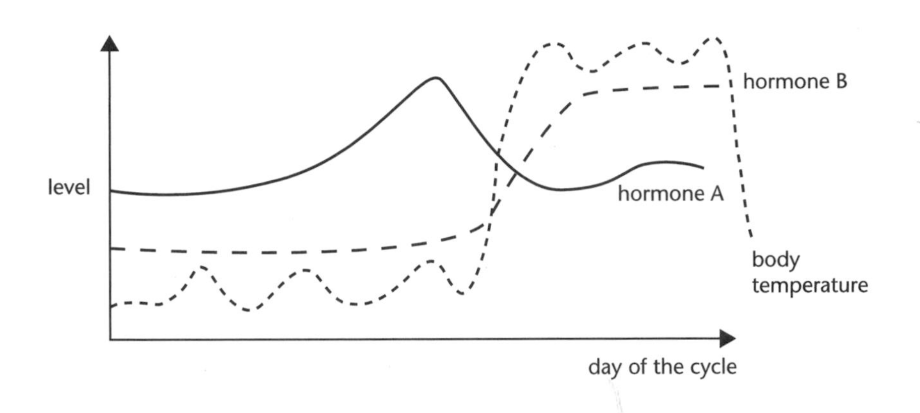

(a) Copy out the following table and complete the blank boxes.

Hormone	Name of hormone	In which half of the cycle does the hormone reach its highest level?
Hormone A		
Hormone B		

[4]

(b) Write down the name of the organ where hormones A and B are made. **[1]**

(c) Describe how the body temperature changes during the cycle. **[2]**

(d) Some women measure their body temperature very accurately. They then use this to prevent getting pregnant or to help them get pregnant.

(i) How could this information help them to get pregnant?

(ii) How could it help to prevent them getting pregnant? **[3]**

(e) Suggest why using body temperature measurements is not a good method on its own to prevent pregnancy. **[1]**

Chapter 2 Health and disease

The following topics are covered in this chapter:

- *Keeping healthy*
- *Giving the body a helping hand*
- *Drugs and health*
- *Too much or too little*

2.1 Keeping healthy

Causes of disease

OCR B B1a
AQA B1.11.4

A disease occurs when the normal functioning of the body is disturbed. **Infectious diseases** can be passed on from one person to another but **non-infectious diseases** cannot.

Genetic diseases are covered on page 44.

Type of disease	Description	Examples
Non-infectious		
• body disorder	incorrect functioning of a particular organ	diabetes, cancer
• deficiency disease	lack of a mineral or vitamin	anaemia, scurvy
• genetic disease	caused by a defective gene	red–green colour blindness
Infectious disease	caused by a pathogen	TB

KEY POINT **Organisms that cause infectious diseases are called pathogens.**

A number of different types of organisms can be pathogens.

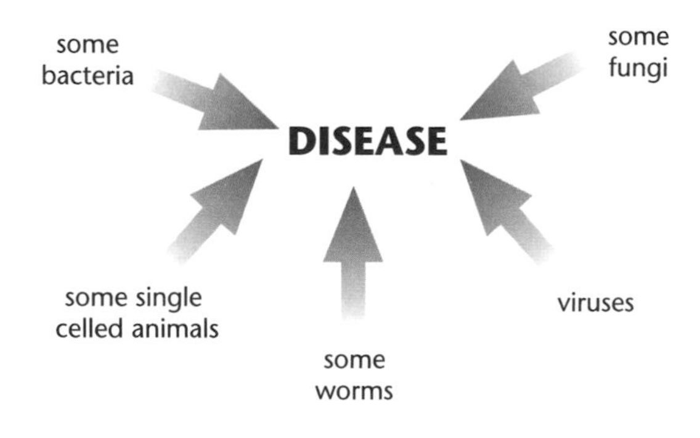

Fig. 2.1 Pathogenic organisms.

Pathogens may reproduce rapidly in the body, either damaging cells directly or producing chemicals called toxins, which make us feel ill.

Viruses damage cells by taking over the cell and reproducing inside them.

How the body protects itself

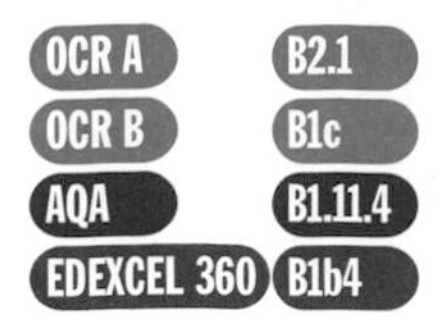

The skin covers most of the body and is quite good at stopping pathogens entering the body.

The body has a number of other defences that it uses in order to try to stop pathogens entering.

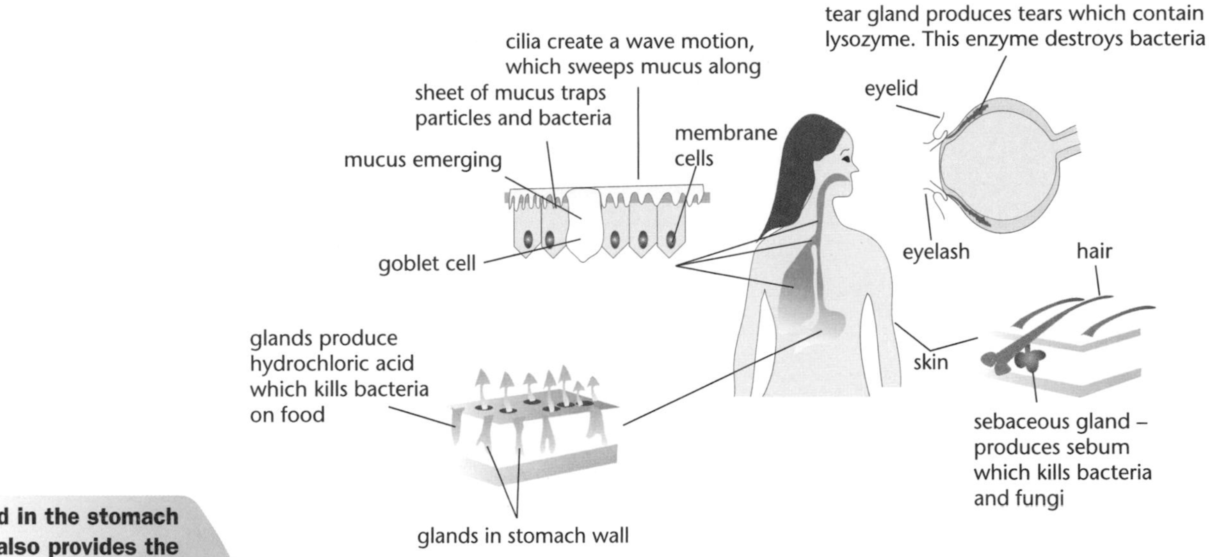

Fig. 2.2 The body's defences.

Acid in the stomach also provides the best pH for protease enzymes to work.

If the pathogens do enter the body then the body will attack them in a number of ways.

The area that is infected will often become **inflamed** and two types of **white blood cells** attack the pathogen.

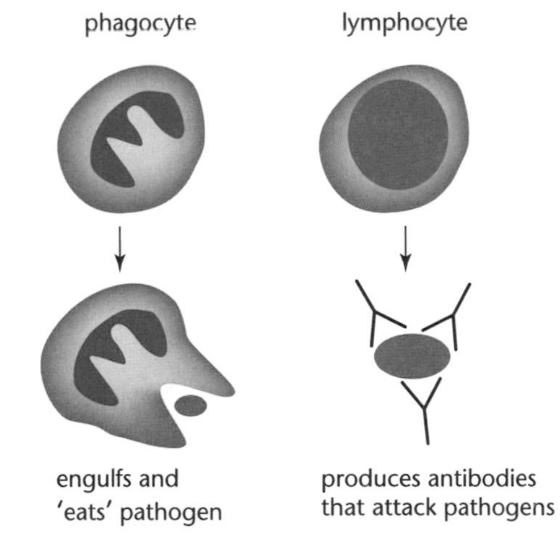

Fig. 2.3 The actions of white blood cells.

The pathogen is detected by the white blood cells because it has foreign chemical groups called **antigens** on its surface. The **antibodies** that are produced by the body are specific to a particular pathogen or toxin and will only destroy that particular antigen.

2.2 Giving the body a helping hand

Antibiotics and antiseptics

OCR A B2.2/3
OCR B B1c
AQA B1.11.4
EDEXCEL 360 B1b4

Antibiotics

Sometimes a pathogen can make us ill before our body's immune system can destroy it. We may sometimes need to take drugs called **antibiotics** to kill the pathogen.

Antibiotics are chemicals that are usually produced by microorganisms that kill bacteria and fungi. They do not have any effect on viruses.

The first antibiotic to be widely used was **penicillin**. Today there are a number of different antibiotics that are used to treat different bacteria. This has meant that some diseases that once killed millions of people can now be treated.

This is an example of natural selection which is covered on page 49.

There is a problem, however. More and more strains of bacteria are appearing that are **resistant to antibiotics**.

A genetic change, or **mutation**, in the bacteria population can enable a large population of resistant bacteria to appear.

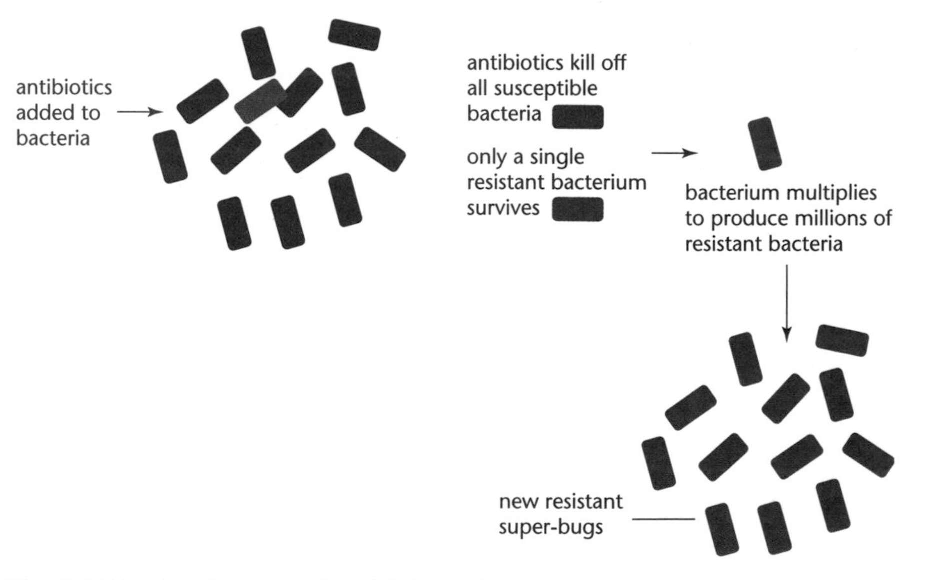

Fig. 2.4 The development of antibiotic resistance.

This process has occurred in many different types of bacteria, including the bacterium causing TB. These bacteria are now resistant to many different types of antibiotic and so are very difficult to treat.

There are various ways that doctors try to prevent the spread of these resistant bacteria:

> *"I tell my patients to finish the dose of antibiotics even if they feel better."*

> *"I prescribe antibiotics only in serious cases caused by bacteria."*

> *"I change the antibiotics that I prescribe regularly and sometimes use combinations of different antibiotics."*

> *"I always wash my hands with antiseptic between patients."*

Antiseptics

One important weapon against resistant bacteria is the use of **antiseptics**.

Antiseptics are artificial chemicals that kill pathogens outside the body.

An antiseptic is usually used on the body and a disinfectant is usually used on other surfaces.

They were first used by an Austrian doctor called Dr Semmelweis to sterilise medical instruments. The use of antiseptics in hospitals is vital in preventing the spread of resistant bacteria.

Vaccinations

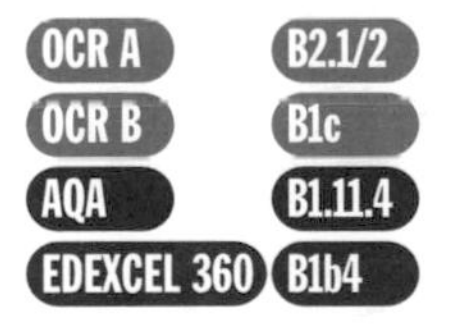

How do vaccines work?

When our body encounters a pathogen, white blood cells make antibodies against the pathogen. If they encounter the same pathogen again in the future then antibodies are produced faster and the pathogen is killed more quickly. This is called **immunity**.

This idea has been used in **vaccinations**.

An immunisation is another name for a vaccination.

A vaccination stimulates our body to make antibodies and special cells called **memory cells**. When the real pathogen comes along, it will be destroyed quickly.

This type of immunity where the antibodies are made by the person is called **active immunity.**

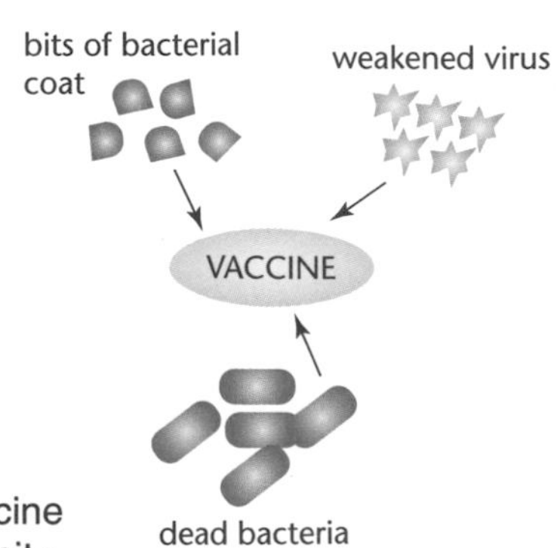

Fig. 2.5 Types of vaccine producing active immunity.

Sometimes it might be too late to give somebody this type of vaccination because they already have the pathogen. They can be given an injection containing antibodies made by another person or animal.

A baby can also get antibodies from breast milk.

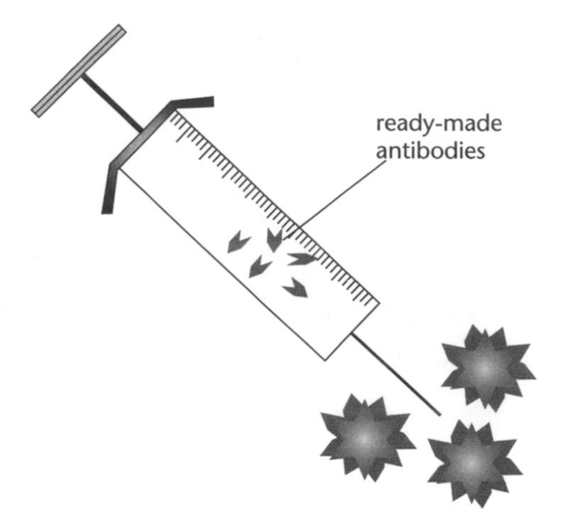

Fig. 2.6 A vaccination containing antibodies.

This is called **passive immunity**. A similar thing happens when a baby receives antibodies from its mother across the placenta.

How long do vaccines protect us for?

Some vaccines protect us for a long time because the memory cells survive for many years. The problem is that some diseases such as influenza need new vaccinations every year.

This is because the virus mutates and changes the shape of its outer coat. This means that different antibodies are needed, so a different vaccination is required.

The human immunodeficiency virus (HIV) also mutates regularly and weakens the immune system. This is making it very difficult to produce a vaccine.

For a vaccine to reduce or completely get rid of a disease, most of the population must be treated. This has sometimes proved difficult as some people are worried about the side effects of certain vaccines.

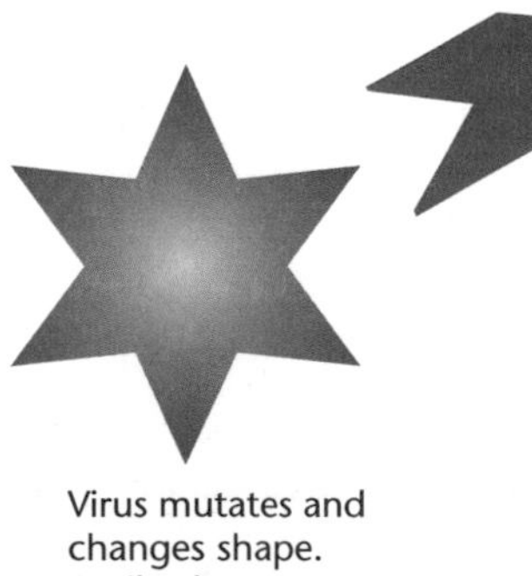

Fig. 2.7 Producing a vaccine.

2.3 Drugs and health

Types of drugs

OCR B B1e
AQA B1.11.3
EDEXCEL 360 B1b4

Drugs are chemicals that alter the functioning of the body.

Some drugs such as antibiotics are often beneficial to our body if used correctly. Others can be harmful, particularly those that are used recreationally.

Sedatives and stimulants often affect the action of synapses (see page 15.)

Many drugs are **addictive**. This means that people want to carry on using them even though they may be having harmful effects. If they stop taking them they may suffer from unpleasant side effects called **withdrawal symptoms**. It also means that people develop **tolerance** to the drug, which means that they need to take bigger doses to have the same effect. Heroin and cocaine are very addictive.

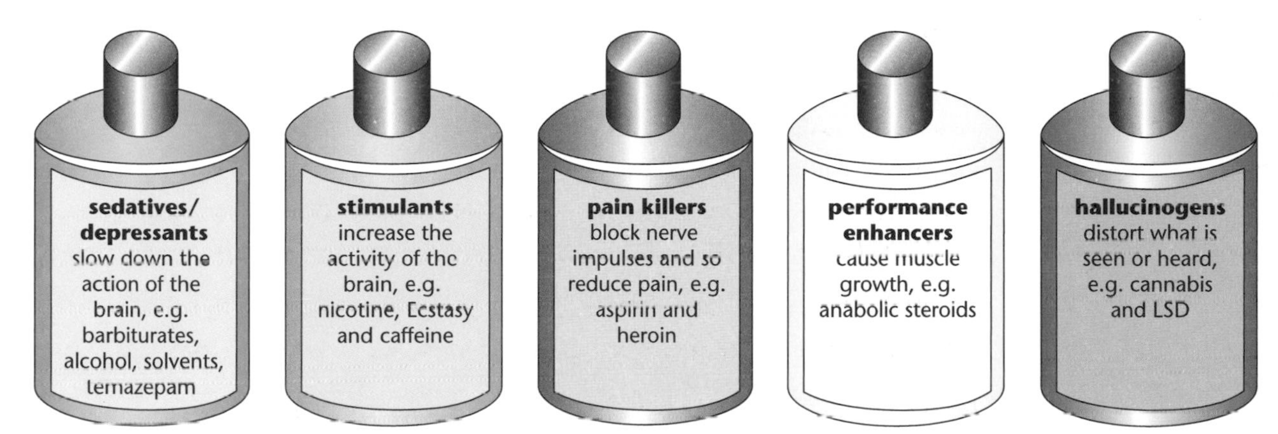

Fig. 2.8 Different drugs do different things.

In order to control drugs, many can only be bought with a prescription. Illegal drugs are classified into groups. Class A drugs are the most dangerous, and class C are the least dangerous. If people are caught with illegal class A drugs the penalties are the highest.

Smoking and drinking alcohol

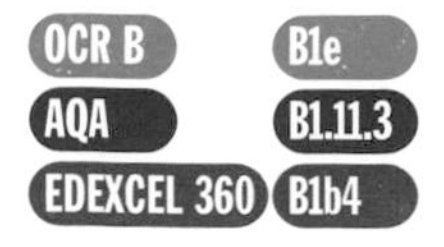

Smoking

Many people cannot give up smoking **tobacco** because it contains the drug **nicotine**. This is addictive. The nicotine is harmful to the body but most damage is done by the other chemicals in the tobacco smoke.

- Chemicals in the tar may cause cells in the lung to divide uncontrollably. This can cause **lung cancer**.
- The mucus collects in the alveoli and may become infected. This may lead to the walls of the alveoli being damaged. This reduces gaseous exchange and is called **emphysema**.
- The heat and chemicals in the smoke destroy the cilia on the cells lining the airways. The goblet cells also produce more mucus than normal. The bronchioles may become infected. This is called **bronchitis**.
- The nicotine can cause an increase in blood pressure increasing the chance of a **heart attack**.

Fig. 2.9 Problems resulting from smoking.

Unlike oxygen, carbon monoxide does not let go of haemoglobin very easily.

Smoking tobacco is particularly dangerous for pregnant women. The **carbon monoxide** in the smoke combines with oxygen in the mother's blood and this can deprive the baby of oxygen. This may lead to a low birth weight.

Drinking alcohol

Drinking alcohol can have a number of effects on the body:

Short-term effects	Long-term effects
upsets balance and muscle control blurred vision and speech slower reactions helps people relax	damage to the liver (cirrhosis) brain damage

Owing to the effects of alcohol on the body there is a legal limit for the level of alcohol in the blood of drivers and pilots.

2.4 Too much or too little

Diseases of excess

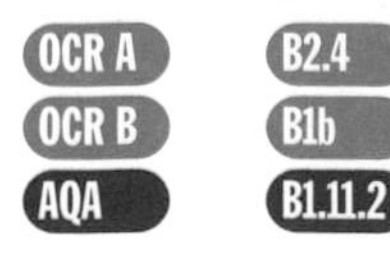

It is important to maintain a balanced diet for the healthy functioning of the body. In the developed world many people eat too much food. This can make a person more likely to get various diseases.

Obesity

If a person eats food faster than it is used up by the body then the excess will be stored. Much of this will be stored as fat and can lead to **obesity**.

Scientists think that up to a million people in the UK might have diabetes now and many do not know they have it.

Obesity can be linked to a number of different health risks:

- **arthritis** – the joints wear out
- **diabetes** – unable to control the blood sugar level
- **breast cancer**
- **high blood pressure**
- **heart disease**.

Doctors take blood pressure using an inflatable cuff around the arm.

It is possible to estimate if a person is underweight, normal, overweight or obese by using the formula:

$$\textbf{Body Mass Index (BMI)} = \frac{\textbf{mass in kg}}{\textbf{(height in metres)}^2}$$

The BMI figure can then be checked in a table to see what range a person is in.

Blood pressure

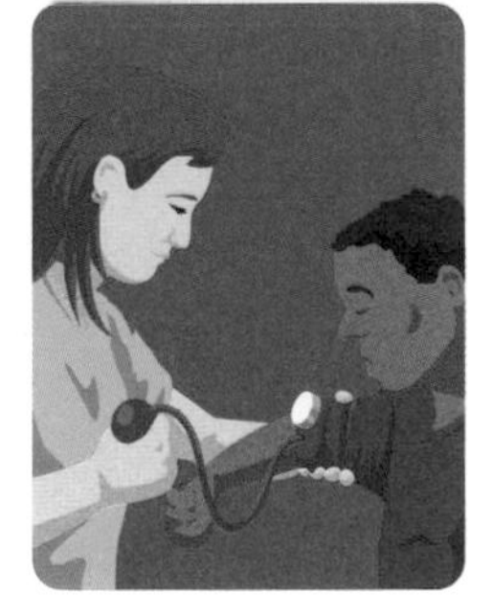

Contractions of the heart pump blood out into the arteries under pressure. This is so it can reach all parts of the body.

Doctors often measure the **blood pressure** in the arteries and give two figures, for example 120 over 80. The highest figure is called the **systolic pressure** and is the pressure when the heart contracts. The second figure is when the heart relaxes and is called the **diastolic pressure**.

Blood pressure varies depending on various factors. The following factors can increase blood pressure:

- high salt and fat in the diet
- lack of exercise
- high alcohol intake
- stress
- obesity
- ageing.

A blood pressure that is too high or too low can cause problems in the body:

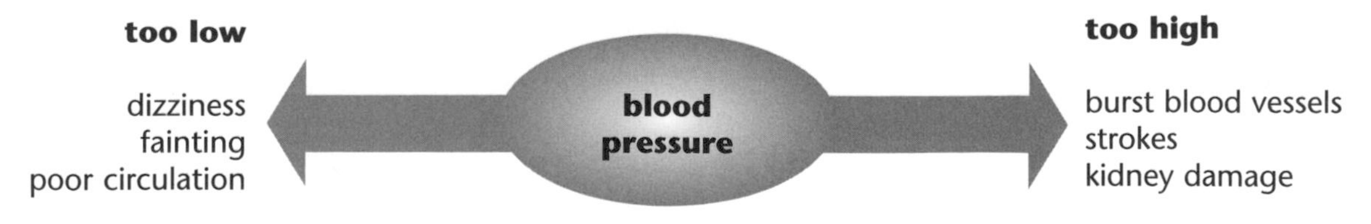

Heart disease

> **Angina is a pain in the shoulders and arm caused by a lack of blood reaching the heart muscle.**

The heart is made up of muscle cells that need to contract throughout life. This needs a steady supply of **energy** so the cells need **oxygen** and **glucose** at all times for **respiration**. This is supplied by blood vessels.

Fatty deposits can form in these blood vessels and reduce the flow of oxygen and glucose to the heart muscle cells.

> **KEY POINT** **This reduction in blood flow causes heart disease and if an area of muscle stops beating then this is a heart attack.**

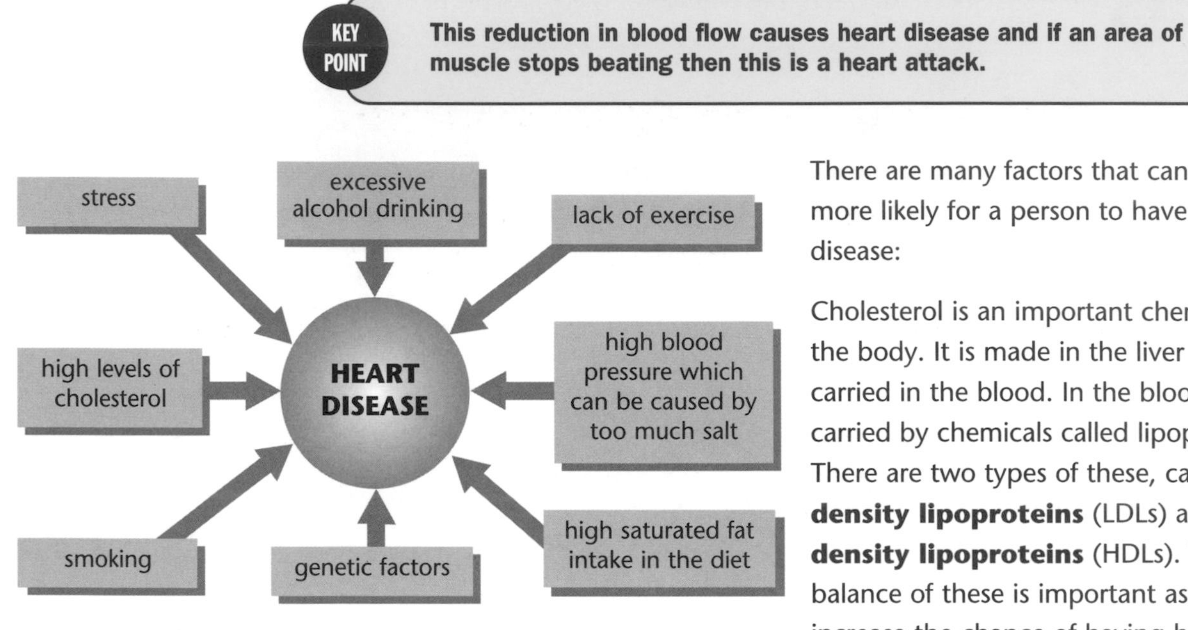

There are many factors that can make it more likely for a person to have heart disease:

Cholesterol is an important chemical in the body. It is made in the liver and carried in the blood. In the blood it is carried by chemicals called lipoproteins. There are two types of these, called **low-density lipoproteins** (LDLs) and **high-density lipoproteins** (HDLs). The balance of these is important as LDLs can increase the chance of having heart disease and HDLs can protect the heart.

Saturated fats increase LDLs and so are bad for you, but polyunsaturated fats increase HDLs.

Deficiency diseases

OCR B B1b, B1c
AQA B1.11.2

In the developing world many people cannot get a balanced diet because there is not enough food to eat.

> **KEY POINT** **Eating too little of one type of food substance can lead to a deficiency disease.**

Examples of **deficiency diseases** are:

- **anaemia** due to a lack of iron
- **scurvy** due to a lack of vitamin C
- **kwashiorkor** due to a lack of protein.

There are times when people do not eat enough food although there is food available. They may put themselves on a diet because they have a poor self-image or think that they are overweight when they are not. This can reduce their resistance to infection and cause irregular periods in women. It may lead to illnesses such as **anorexia**.

HOW SCIENCE WORKS

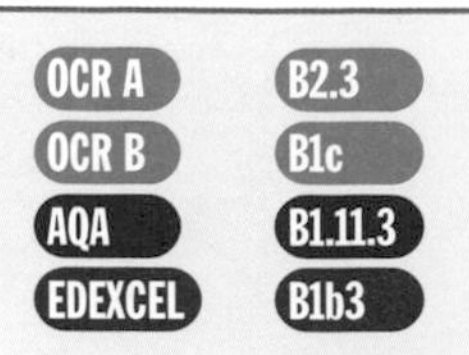

Drug testing – an ethical problem

New drug makes men dangerously ill

In March 2006, six men volunteered to take part in tests on a new drug. The drug had been designed to treat serious diseases such as rheumatoid arthritis, leukaemia and multiple sclerosis.

This was the first time the drug had been used on people and unfortunately it caused serious side effects on the men, who became very ill. But why was this drug tested on people and had it been tested elsewhere first?

- Once a new drug has been made, it is tested on cells in a laboratory.
- If it passes these tests it is then tried on animals.
- Then it is tested on human volunteers.
- Finally, it is tested on human patients with the disease.

Many of these tests cause disagreements.

Many people think that animals should not be used to test drugs. Some think that it is too cruel, others think that it is pointless.

"The case in March 2006 shows that it is pointless testing drugs on animals. They do not react the same way as people."

"We must carry on testing on animals. This has prevented thousands of harmful drugs being tested on people."

Once the drug is cleared to be tested on patients the trial has to be set up carefully. One group are given the drug and another group have a **placebo**. This looks like the real treatment but has no drug in it.

The two groups do not know which treatment they are having, nor does the doctor who treats them. This is called a **double blind test** and it means that the people are not influenced by knowing which treatment is being given.

There are many ethical issues involved in testing drugs.

Some people think that tests are cruel or dangerous and therefore should not be done.

Others think that the tests are reasonable because the **benefits outweigh the risks** of the tests.

HOW SCIENCE WORKS

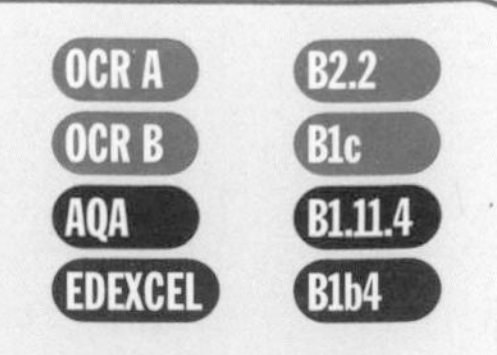

The risks of vaccinations

Whether or not to give your children vaccinations is a difficult decision for some people to make. Diseases like measles mumps and rubella can have serious effects on the body.

- **Measles** is a very serious disease. 1 in 2500 babies that catch the disease die.
- **Mumps** may cause deafness in young children. It may also cause viral meningitis which can be fatal.
- **Rubella** can cause a baby to have brain damage if its mother catches the disease during pregnancy.

The graph shows the number of people getting measles in one country (Ireland) each year.

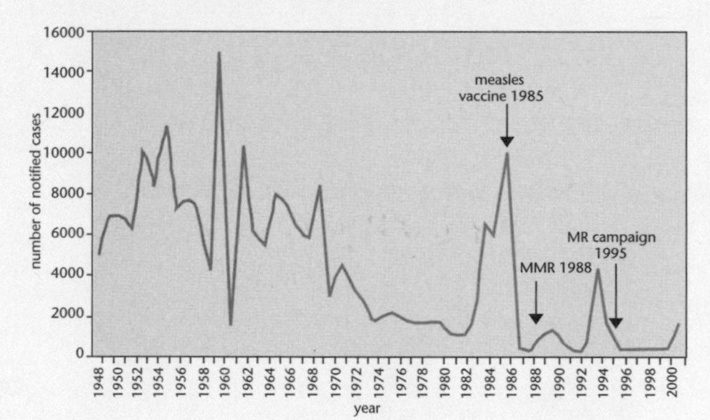

A measles vaccine was introduced in 1985.

In 1988, a combined measles, mumps and rubella vaccine was introduced.

A measles and rubella (MR) vaccination campaign for primary school-age children was conducted in 1995.

The introduction of the measles vaccine and the MMR vaccine has led to a decrease in measles. However, the uptake of MMR was not high enough to prevent outbreaks in 1993 and 2000.

Over 1600 cases of measles were notified in Ireland in 2000. Measles has been eliminated or is close to elimination in Finland, Spain and other European countries where there is good uptake of vaccines. In Ireland, 8 deaths from measles were reported between 1990 and 1999 and in 2000, three children in Dublin died from the disease.

So why are people worried about the vaccination?

In 1998, a study of autistic children raised the question of a connection between MMR vaccine and autism. (People with autism have difficulty with communicating and using some thinking skills.)

The 1998 study has a number of limitations. For example, the study was very small, involving only 12 children. This is too few cases to make any generalisations about the causes of autism. In addition, the researchers suggested that the MMR vaccination caused bowel problems in the children, which then led to autism. However, in some of the children studied, symptoms of autism appeared before symptoms of bowel disease.

Some people say that parents should be forced to have their children vaccinated otherwise the disease will not disappear.

It is impossible to say that having a vaccine does not involve a **risk**. However, like many things in life, it is a question of balancing risks, the risks of the vaccination against the risk of the disease.

HOW SCIENCE WORKS *Questions*

1. Suggest two reasons why drugs are tested on animals. **[2]**
2. Explain why placebos are used when testing drugs. **[2]**
3. In the 1960s, a drug called thalidomide was given to pregnant women. The drug had been tested on animals and seemed to be safe. Soon it became clear that the drug was causing the women to give birth to children with birth defects. Some animal rights campaigners say that this shows that drug testing on animals is a waste of time. Write down an argument for and against this view. **[4]**

Exam practice questions

1. Which part of the blood produces antibodies?

A red blood cells
B white blood cells
C platelets
D plasma **[1]**

2. Which of the following could increase the risk of getting heart disease?

A more stress
B less saturated fat in the diet
C stopping smoking
D drinking a glass of wine a day **[1]**

3. Lack of vitamin C in the diet could cause

A kwashiorkor
B anorexia
C anaemia
D scurvy **[1]**

4. Fill in the gaps in the following sentences using words from the list:

acid antibodies lysozyme pathogen sebum toxins

An organism that causes a disease is called a ____1____. The body tries to prevent these organisms entering the body. Tears contain ____2____ and the stomach makes ____3____, both of which can destroy the organisms.

If the organism enters the body it can damage cells or release poisons called ____4____. **[4]**

5. The MMR vaccination is usually given to young children. It protects them from measles, mumps and rubella. Here is some information about these three diseases.

- **Measles** is a very serious disease. 1 in 2500 babies that catch the disease die.
- **Mumps** can cause deafness in young children.
- **Rubella** can cause a baby to have brain damage if its mother catches the disease during pregnancy.

(a) The MMR vaccination is given to young children rather than waiting until they are older. Write down **one** reason why. **[1]**

(b) Pregnant women are tested to see if they have had the MMR vaccination. Why is this important? **[1]**

(c) Explain how a vaccination such as MMR can protect a person from getting a disease. **[3]**

(d) People often feel ill after having a vaccination like MMR. Explain why this is. **[2]**

Exam practice questions

6. The following table gives some information about four drugs.

Drug	Type of action	Addictive?
anabolic steroids	performance enhancer	no
aspirin	pain killer	no
barbiturate	depressant	yes
nicotine	stimulant	yes

(a) Which drug shown in the table is found in cigarette smoke? **[1]**

(b) Two of the drugs in the table are addictive. What does this mean? **[2]**

(c) What effect would barbiturates have on the nervous system? **[1]**

(d) Barbiturates are a class B drug. Why are drugs put into different classes? **[2]**

Chapter

Variation and genetics

The following topics are covered in this chapter:

- ***Genes and chromosomes***
- ***Manipulating genes***
- ***Variation and evolution***

3.1 Genes and chromosomes

What is a gene?

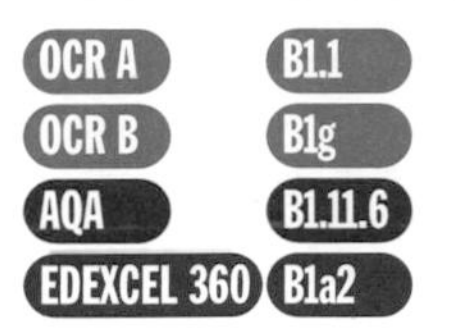

Most cells contain a nucleus that controls all of the chemical reactions that go on in the cell. Nuclei can do this because they contain the **genetic material**. Genetic material controls the characteristics of an organism and is passed on from one generation to the next. The genetic material is made up of structures called **chromosomes**. They are made up of a chemical called deoxyribonucleic acid or **DNA**. The DNA controls the cell by coding for the making of proteins, such as enzymes. The enzymes control all the chemical reactions taking place in the cell.

A gene is part of a chromosome that codes for one particular protein.

DNA codes for the proteins it makes by the order of four chemicals called bases. They are given the letters **A**, **C**, **G** and **T**.

Each one of our cells has about 30 000 genes.

Although the cells in one person have the same genes, they do not use them all. For example, a liver cell will use some genes and a cheek cell will use different genes.

The Human Genome Project has mapped all the genes on our chromosomes.

By controlling cells, genes therefore control all the characteristics of an organism. Different organisms have different numbers of genes and different numbers of chromosomes. In most organisms that reproduce by sexual reproduction, the chromosomes can be arranged in pairs. This is because one of each pair comes from each parent.

Chromosomes and reproduction

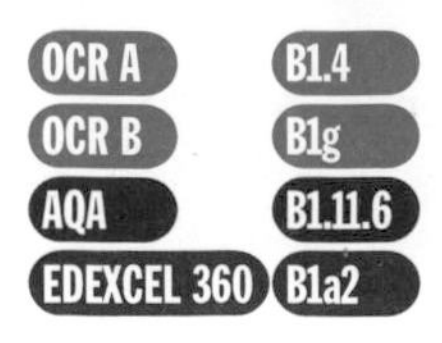

No living organism can live for ever, so there is a need to reproduce. There are two main methods that organisms use to reproduce: sexual and asexual.

Sexual reproduction

Sexual reproduction involves the passing on of genes from two parents to the offspring. This is why we often look a little like both of our parents. The genes are passed on in the sex cell or gametes.

KEY POINT **Sexual reproduction involves the joining together of male and female sex cells or gametes.**

Identical twins are made when the embryo splits into two, so they have the same genes as each other.

In humans each body cell has 46 chromosomes in 23 pairs. This means that when the male sex cells (sperm) are made they need to have 23 chromosomes one from each pair. The female gametes (eggs) also need 23 chromosomes. When they join at fertilisation they produce a cell called a **zygote** that has 46 chromosomes again. This will grow into an **embryo** and become a baby.

This also means that the offspring that are produced from sexual reproduction are all different because they have different combinations of chromosomes from their mother and father.

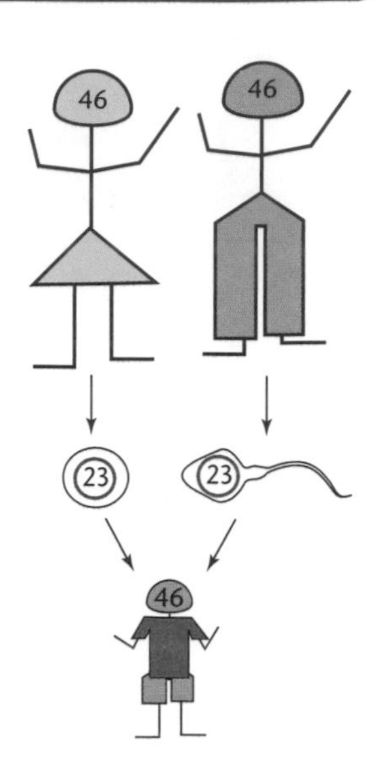

Fig. 3.1 Sexual reproduction.

Asexual reproduction

Gardeners often use asexual reproduction to copy plants – they know what the offspring will look like.

Bacteria, plants and some animals can reproduce asexually. This needs only one parent and does not involve sex cells joining.
All the offspring that are made are genetically identical to the parent.
Different organisms have different ways of reproducing asexually but one example is the spider plant. This grows new plantlets on the end of long shoots.

Fig. 3.2 A spider plant uses asexual reproduction to reproduce.

Sex determination

OCR A B1.2
OCR B B1h

Humans have 23 pairs of **chromosomes**. The chromosomes of one of these pairs are called the sex chromosomes because they carry the genes that determine the sex of the person.

KEY POINT **There are two kinds of sex chromosome. One is called X and one is called Y.**

- Females have **two X chromosomes** and are **XX**.
- Males have an **X and a Y chromosome** and are **XY**.

It is possible to separate X and Y sperm in the laboratory and so choose the sex of a baby.

This means that females produce ova that contain single X chromosomes. Males produce sperm, half of which contain a Y chromosome and half of which contain an X chromosome.

The reason why the sex chromosomes determine the sex of a person is due to a single gene on the Y chromosome. This gene causes the production of testes rather than ovaries and so the male sex hormone testosterone is made.

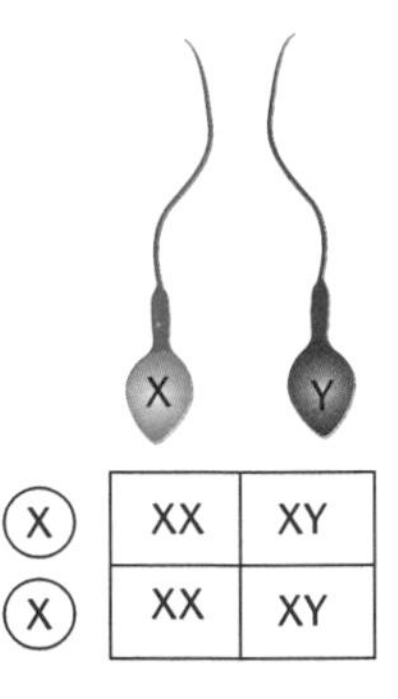

Passing on genes

Because we have two copies of each chromosome in our cells (one from each parent) this means that we have two copies of each gene.

A copy of a gene is called an allele.

Sometimes the two alleles are the same but sometimes they are different. A good example of this is tongue rolling. This is controlled by a single gene and there are two alleles of the gene, one that says roll and the other that says do not roll.

The people who are non-rollers must have two recessive alleles.

But even if we have two alleles, if one says roll and the other says do not roll, then a person can still roll their tongue. This is because the allele for rolling is **dominant** and the non-rolling allele is **recessive**.

KEY POINT

Some important words that you need to know:
- **homozygous means that both alleles are the same**
- **heterozygous means that the two alleles are different.**

How to work out the results of a cross

We usually give the alleles letters, with the dominant allele having a capital letter. For example, T = tongue rolling and t = non-rolling.

Let us assume that mum cannot roll her tongue but dad can.

Both of dad's alleles are T so he is homozygous.

The cross is usually drawn out like this:

In this cross all the children can roll their tongue.

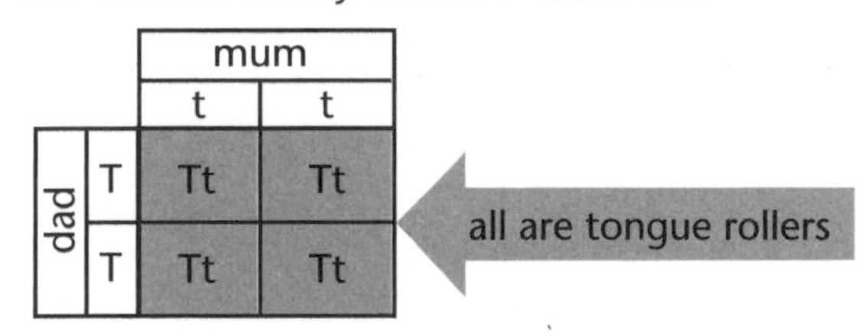

In this cross, 1 in 4 or 25% or a quarter of the children cannot roll their tongue.

If both mum and dad are heterozygous the children that they can produce will be different:

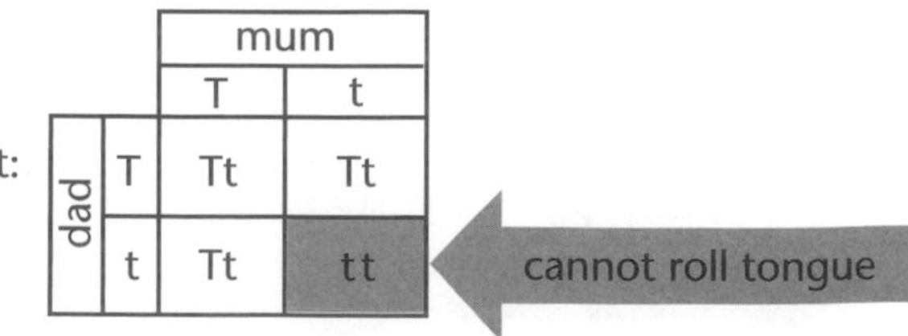

Inherited disorders

Many genetic disorders are caused by certain alleles. These can be passed on from mother or father to the baby and lead to the baby having the disorder. Examples of these disorders are **cystic fibrosis** and **Huntington's disorder**.

Cystic fibrosis is caused by a recessive allele and Huntington's is caused by a dominant allele.

People with these disorders become ill:

Cystic fibrosis	Huntington's disorder
mucus collects in the lung	muscle twitching
breathing is difficult	loss of memory
food is not properly digested	difficulty in controlling movements

Remember genetic cross diagrams can only work out the probability of a child being affected.

By looking at family trees of these genetic disorders and drawing genetic diagrams like the one for tongue rolling, it is possible for people to know the chance of them having a child with a genetic disorder. This may leave them with a difficult decision to make as to whether to have children or not.

3.2 Manipulating genes

Genetic engineering

All living organisms use the same language of DNA. The four 'letters' **A**, **G**, **C** and **T** are the same in all living things. Therefore a gene from one organism can be removed and placed in a totally different organism where it will continue to carry out its function.

KEY POINT **Moving a gene from one organism to another is called genetic engineering.**

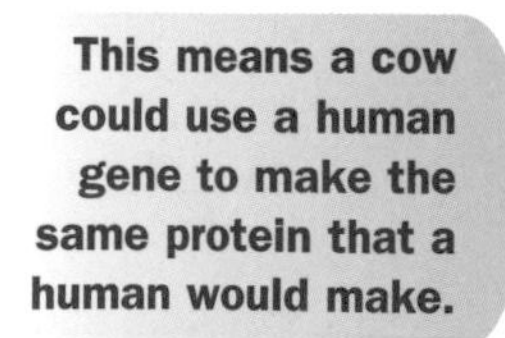

This means a cow could use a human gene to make the same protein that a human would make.

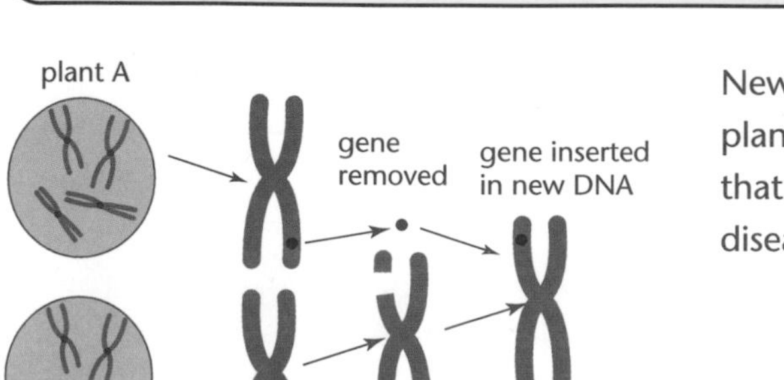

New genetically modified (GM) plants can be made in this way that may be more resistant to disease or produce a higher yield.

Fig. 3.3 Genetic engineering.

People often have different views about genetic engineering:

- Some people think that genetic engineering is against 'God and Nature' and is potentially dangerous.
- Some people think that genetic engineering will provide massive benefits to humans, such as better food and less disease.

There is also the possibility that genetic engineering may be used to treat genetic disorders like cystic fibrosis. Scientists are trying to replace the genes in people who have the disorder with working genes.

Using genetic engineering to treat genetic disorders is called gene therapy.

Again this is quite controversial because it could be used to change the genes of embryos. People are worried that it might be used to produce 'designer babies'.

Cloning

OCR A B1.4
AQA B1.11.6
EDEXCEL 360 B1a2

Asexual reproduction produces organisms that have the same genes as the parent.

This means that they will be very similar.

Genetically identical individuals are called clones.

Many plants such as the spider plant do this naturally and it is easy for a gardener to **take cuttings** to make identical plants.

Modern methods involve **tissue culture**, which uses small groups of cells taken from plants to grow new plants.

Cloning animals is much harder to do. Two main methods are used:

- **Cloning embryos**. Embryos are split up at an early stage and the cells are put into host mothers to grow.
- **Cloning adult cells**. The first mammal to be cloned from adult cells was Dolly the sheep.

Since Dolly was born other animals have been cloned and there has been much interest about cloning humans.

There could be two possible reasons for cloning humans:

- **Reproductive cloning** to make embryos for infertile couples.
- **Therapeutic cloning** to produce embryos that can be used to treat diseases.

The use of embryos to treat disease is possible due to the discovery of stem cells.

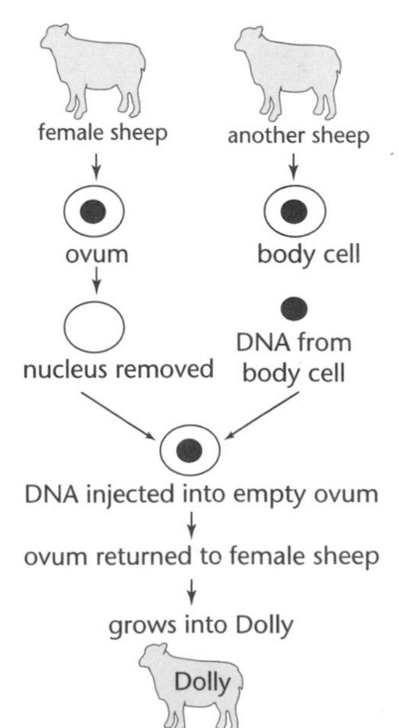

Fig. 3.4 How Dolly was cloned.

Stem cells are cells that have the ability to divide to make any of the different tissues in the body.

It is much easier to find stem cells in an embryo and scientists think that they could be used to repair damage such as injuries to the spinal cord.
There are therefore many different views about cloning:

“Both infertility and genetic diseases cause much pain and distress. I think that we should be able to use cloning to treat these problems.”

“It is not right to clone people because clones are not true individuals and it is not right to destroy embryos to supply stem cells”

3.3 Variation and evolution

Variation

OCR A B3.1
OCR B B1h

Children born from the same parents all look slightly different. We call these differences ‘variation’:

- **Inherited or genetic** – some variation is inherited from our parents in our genes.
- **Environmental** – some variation is a result of our environment.

Often our characteristics are a result of both our genes and our environment. Here are some examples:

Inherited	Environmental	Inherited and environmental
earlobe shape	scars	intelligence
eye colour	spoken language	body mass
nose shape		height

Rare cases of identical twins that have been brought up separately can provide good evidence to investigate this.

The genes provide a height and weight range into which we will fit, and how much we eat determines where in that range we will be.
Scientists have argued for many years whether ‘nature’ or ‘nurture’ (inheritance or environment) is responsible for characteristics such as intelligence, sporting ability and health.

Because the baby can receive any one of the 23 pairs from mum and any one of the 23 pairs from dad, the number of possible gene combinations is enormous. This new mixture of genetic information produces a great deal of variation in the offspring. This just mixes genes up in different combinations but the only way that new genes can be made is by **mutation**.

A mutation is a random change in a gene.

A gene mutation occurs when one of the chemical 'letters' in DNA is changed. When this happens, it is most unlikely to benefit the organism. Think what would happen if you made random changes to a few of the letters on this page. It is most likely to produce gibberish and very unlikely to make any sense at all.

- If a mutation occurs in a gamete, the offspring may develop abnormally and could pass the mutation on to their own offspring.
- If a mutation occurs in a body cell, it could start to multiply out of control – this is **cancer**.

But very occasionally, a mutation may be useful, and without mutations we would not be here.

This shows how important it is to wear sun protection cream when in strong sunlight.

Here are some causes of mutations:

- **radiation**
- **UV** in sunlight
- **X-rays**
- **chemical mutagens** – as found in cigarettes.

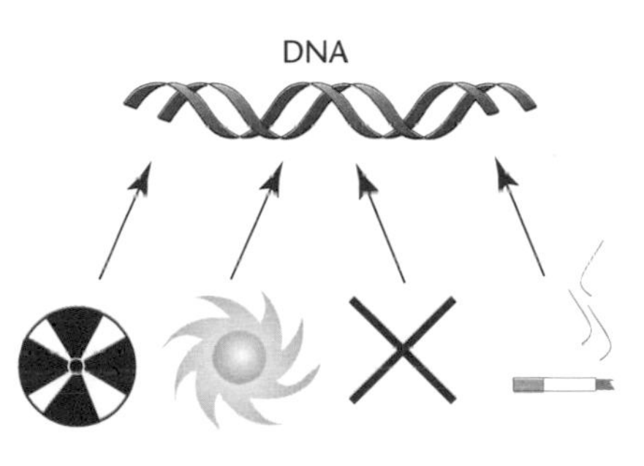

Fig. 3.5 Causes of mutations.

Evolution

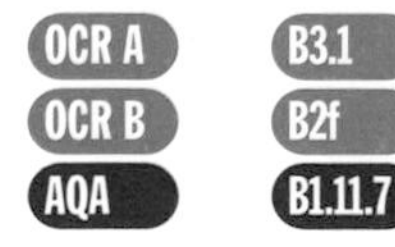

Most scientists now think that life on Earth started about 3500 million years ago. But how life started and why there is such a great variety of organisms have been questions that people have argued over for a long time.

In the 1800s, scientists started to find **fossils** of many different animals and plants.

Records of organisms can be preserved in different ways:

hard body parts get covered in sediment and gradually harden with minerals to become fossils

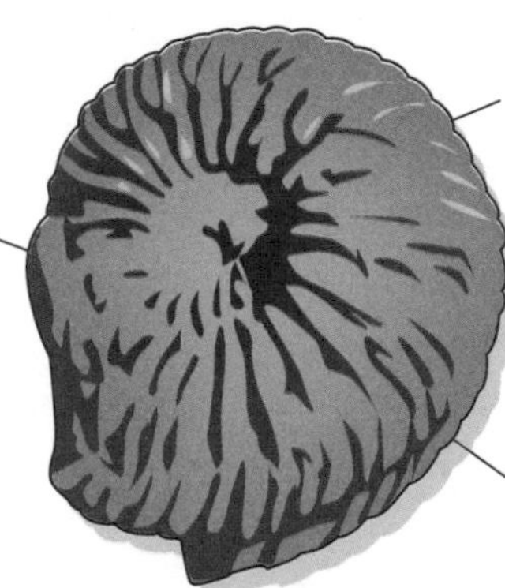

casts or impressions can be found, such as footprints

specimens can be found in amber, peat bogs, tar pits and ice

Fig. 3.6 Evidence from dead organisms.

Many people at that time believed in **creation**. They said that organisms were created as they exist now by God. However, scientists found fossils of organisms such as **dinosaurs** that are not alive today. Some people started to believe the idea that species of organisms could gradually change.

KEY POINT

Evolution is the gradual change in a species over a long period of time.

People argued against evolution by saying that the fossil record was incomplete and that the fossils of some organisms are missing.
Possible reasons for this are:

- Some body parts do not fossilise.
- Conditions have to be just right for fossils to be made.
- Many fossils may not yet have been found.

The problem for the believers in evolution was that at first they could not explain how the gradual changes happened.

Charles Darwin and natural selection

OCR A B3.1
OCR B B2f
AQA B1.11.7
EDEXCEL 360 B1a1

Many of these observations were made on the Galapagos Islands off the coast of South America.

Charles Darwin (1809–1882) was a naturalist on board a ship called the HMS *Beagle*. His job was to make a record of the wildlife seen at the places the ship visited.

On his travels, Darwin noticed four things:

- Organisms often produce **large numbers** of offspring.
- Population numbers usually remain constant over long time periods.
- Organisms are all slightly **different**: they show **variation**.
- This variation can be inherited from their parents.

Darwin used these four simple observations to come up with a theory for how evolution could have happened.

Darwin said that all organisms are slightly different and some are **better suited** to the environment than others. These organisms are more likely to survive and reproduce.

They will pass on these characteristics and over long periods of time the species will change.

KEY POINT

Darwin called this theory natural selection.

Darwin was rather worried about publishing his ideas. When he finally published them they caused much controversy. Many people were very religious and believed in creation. It took many years before Darwin's theory was generally accepted.

Natural selection in action

Because natural selection takes a long time to produce changes, it is very difficult to see it happening. One of the first examples to be seen was the peppered moth.

Fig. 3.7 Light and dark peppered moths.

This moth is usually light coloured, but after the Industrial Revolution a black type became common in polluted areas. This can be explained by **natural selection**:

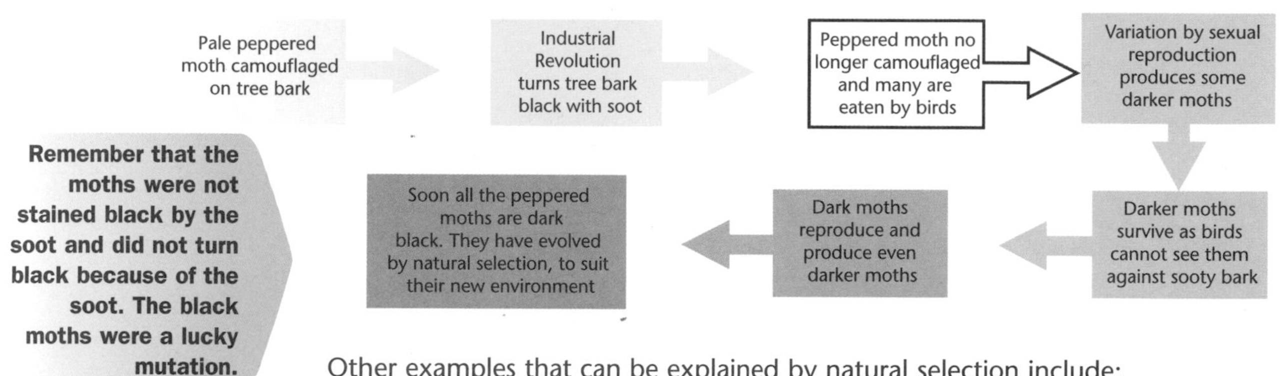

Remember that the moths were not stained black by the soot and did not turn black because of the soot. The black moths were a lucky mutation.

Other examples that can be explained by natural selection include:

- Rats becoming resistant to the rat poison warfarin.
- Bacteria becoming resistant to antibiotics.

MRSA bacteria are discussed on page 29.

How did life start?

OCR A B3.1

The atmosphere at the time contained different gases from those in the atmosphere now. The electric spark copies the action of lightning.

Once people accepted natural selection and evolution, scientists tried to work out how life might have started.

Experiments have shown that molecules that are important to life can be made in conditions that are similar to those found on the Earth 3500 million years ago.

Molecules that make up DNA can be produced and DNA is found in all living cells. It can copy itself and carry genetic codes, so it is vital to life.

Some scientists think that other molecules may have come from space, perhaps from comets.

Once living cells were produced, natural selection could produce the variety of life that is alive today.

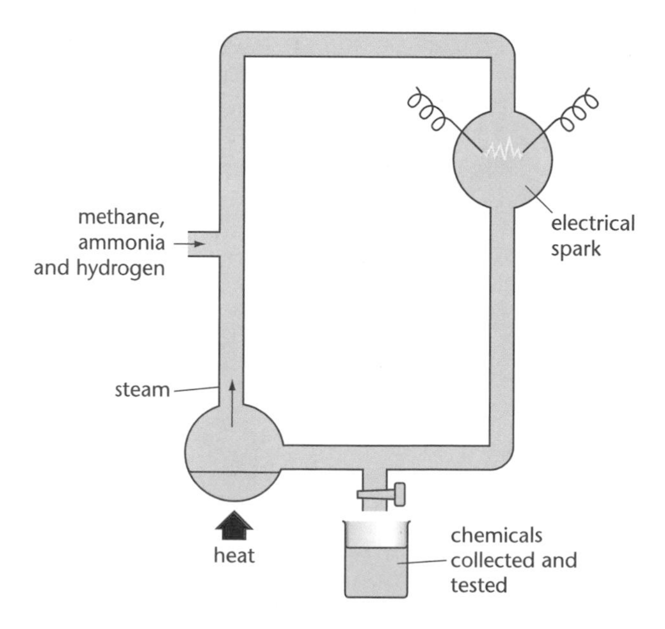

Fig. 3.8 How to create molecules important to life.

HOW SCIENCE WORKS

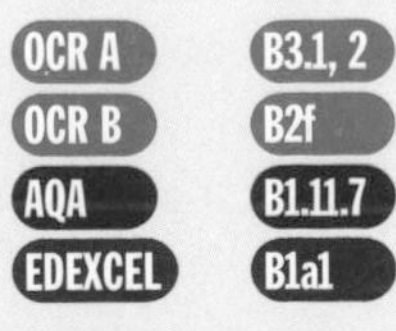

Evolution – explaining the facts

Living on the Earth are about five million different types of living organisms.

During humankind's time on Earth many people have put forward ideas to try and explain how this has come about.

Three different ideas are shown:

Creation theory says that the Earth and life on it were created by God as described in the Bible. Only small changes have happened since creation and no new species have been created.

Darwin said that all organisms were slightly different. Those organisms that were better suited would pass on their characteristics and so the population would gradually change.

A French scientist called **Lamarck** said that organisms were changed by their environment during their life. They then passed on the new characteristics and so the population would change.

Each of these ideas is a **theory** that explains **data** that is known at the time. As different data becomes known, then people often start to accept different theories. Since the discovery of many different extinct fossils and the dating of rock many people think that the creation theory cannot explain the data.

Most scientists now think that Lamarck's theory is wrong because we now know that characteristics are passed on in our genes and genes are not usually altered by the environment.

Most people now accept Darwin's theory because it best explains all the data that has been discovered. But it is only a theory, not fact.

HOW SCIENCE WORKS

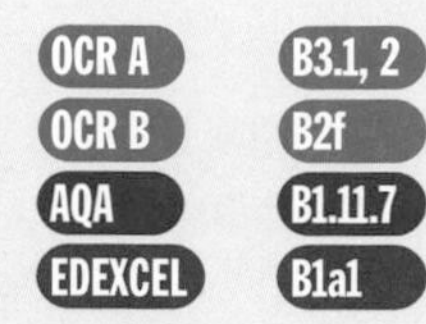

Manipulating genes – right or wrong?

In 1990, two American girls took part in the world's first gene therapy experiment to try to repair their immune systems. The gene therapy involved putting genes into their white blood cells, using a virus to inject the genes. The normal virus genes had first been removed. The experiment was successful. Without those new genes, it was unlikely they could have survived. Their survival signalled the possible start of using gene therapy to cure many life-threatening genetic disorders.

But after 5000 patients had participated in 350 trials, things began to go wrong. First, in 1999, a young man being treated died of a massive immune reaction to the gene treatment. Three years later, a French baby developed leukaemia as a by-product of the treatment. Experts say the virus that inserted the genes mistakenly turned on a cancer-causing gene.

Scientists are now learning from these mistakes and further research into gene therapy is taking place. Cystic fibrosis is the most common inherited disease in the Western world. In 1989, the gene that causes cystic fibrosis was discovered. This caused excitement that this disorder could be treated with gene therapy.

Scientists have tried for ten years to put genes into the cells of the lungs to cure cystic fibrosis, but with little success. It seems to work in cells in a test tube but not in the body.

Now a new idea is about to be tried. This involves using stem cells. These are cells that can develop into different types of cells. They can be taken from the body of the patient and then the normal gene can be put inside them. The cells can then be put back into the patient's lungs where they divide to make normal lung cells.

The possible use of gene therapy and stem cells raises some ethical issues. To make the stem cells involves the destroying of embryos.

Some people think that this is not acceptable. Others think that it is wrong to stop people having a treatment that might cure their disease.

*"I think that destroying embryos is wrong even if it might help to cure people. It is not **ethically acceptable**."*

"I think that it is wrong to stop people having a treatment that might cure their disease."

HOW SCIENCE WORKS Questions

When Darwin returned from his journey on the Beagle he wrote down his ideas about how evolution may have happened.

1. Suggest why Darwin was worried about publishing these ideas. **[2]**
2. How did the discovery of fossils of extinct animals help to persuade some people that Darwin's ideas were right? **[2]**
3. What is the difference between a theory such as natural selection and data such as fossils? **[3]**

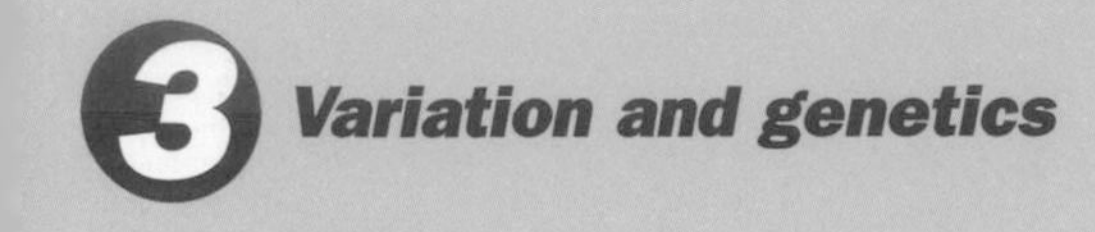

Exam practice questions

1. The theory put forward by Charles Darwin to explain how evolution could have occurred is called:

A selective breeding
B cloning
C natural selection
D genetic engineering **[1]**

2. Which row, A, B, C or D, in the following table contains the correct characteristics? **[2]**

	Characteristic is controlled by		
	Genes only	Environment only	Both
A	height	eye colour	body mass
B	body mass	scars	earlobe shape
C	scars	body mass	height
D	blood group	spoken language	intelligence

3. A genetically identical copy of an individual is called a:

A clone
B mutation
C gamete
D zygote

4. Finish the sentences by writing the correct word in the gap. Choose your answers from this list.

chromosomes **cytoplasm** **DNA** **nucleus** **proteins** **sugar**

The genes in cells are found in the ______________.
These genes are on long strands called ________________ .
They are made of a chemical called ____________.
The genes control the cell by dictating which _____________ the cell should make. **[4]**

5. The diagram shows the chromosomes present in a human skin cell.

Exam practice questions

(a) How would these chromosomes be different if they were from a sex cell and not a skin cell? **[2]**

(b) The cell is from a male. How can you tell this? **[2]**

6. Jackie has just had a baby. Leroy is the father. The doctor has told them that their baby has cystic fibrosis even though neither Jackie nor Leroy has the disorder.

(a) Is the copy of the gene (allele) that causes cystic fibrosis dominant or recessive? How can you tell? **[2]**

(b) Fill in the boxes below to show how Jackie and Leroy could have had a baby that has cystic fibrosis. Jackie's gametes have been done for you. **[3]**

Jackie

	gametes	**F**	**f**
Leroy			

(c) If Jackie and Leroy had another baby, what would be the chance of it having cystic fibrosis? **[1]**

(d) The doctor tells Jackie that if she gets pregnant again they could have the foetus tested to see if it had cystic fibrosis. Suggest why it may be a difficult decision for Jackie to decide whether to be tested or not. **[2]**

Chapter 4

Organisms and the environment

The following topics are covered in this chapter:

- ***The variety of life***
- ***Living together***
- ***Human impact on the environment***
- ***Conservation***

4.1 The variety of life

Classifying organisms

OCR B B1b
EDEXCEL 360 B1a1

Humans have been classifying organisms into groups ever since they started studying them.

- This makes it convenient when trying to identify an unknown organism.
- It also tells us something about how closely organisms are related and about their evolution.

The modern system that we use puts organisms into a system of smaller and smaller groups. **Kingdoms** are the largest groups. The kingdoms are divided into smaller and smaller groups until the smallest group is formed, called a **species**.

Members of a species are very similar, but how do we know if two similar animals are the same species?

KEY POINT **Members of the same species can breed with each other to produce fertile offspring.**

Animals such as mules are called hybrids and are often very strong.

Horses and donkeys are different species because, although they can mate and produce a mule, mules are **infertile**.

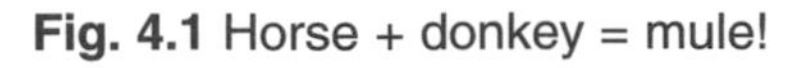

Fig. 4.1 Horse + donkey = mule!

Deciding on groups

The first step in classifying an organism is to put it into a **kingdom**. Two of the kingdoms are the **plant and animal kingdoms**.

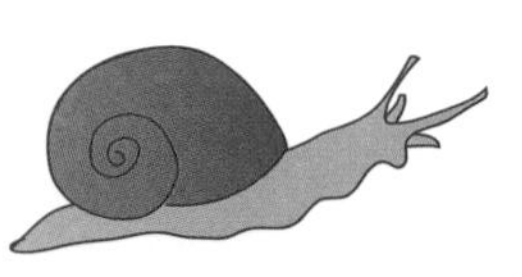

Most of the differences between plants and animals are due to the way that they feed.

Plants	Animals
make their own food	take their food in ready-made
move by growing	move the whole of their body
grow in a spreading shape	have a compact shape

This makes it easy to classify most plants and animals but there are problems. **Fungi** grow rather like plants but cannot make their own food and therefore they are put in a separate kingdom.

Some microscopic organisms like *Euglena* can make their own food like plants but in the dark start feeding like an animal. They are put in another kingdom called **Protoctista**.

Bacteria are so different in structure from other organisms that they are put in a separate kingdom. This makes five kingdoms in all.

Once an organism is put into the animal kingdom it can be put into the vertebrate group or one of several invertebrate groups.

The vertebrate group is divided into five different classes.

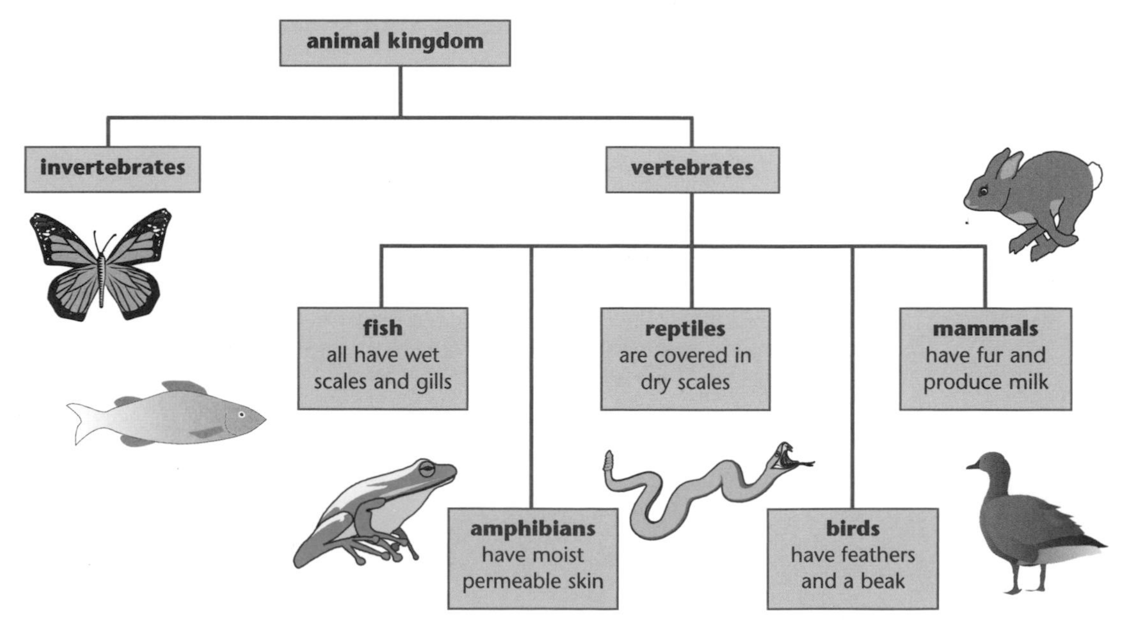

Fig. 4.2 The animal kingdom.

Naming organisms

OCR B B2b

Organisms are often known by different names in different countries or even in different parts of the same country. All organisms are therefore given a **scientific name** that is used by scientists in every country. This avoids confusion.

KEY POINT **The scientific system of naming organisms is called the binomial system.**

Each name has two parts. The first part is the name of the genus (the group above species). The second part of the name is the species. For example:

The binomial name is always typed in italics and the genus starts with a capital letter but the species does not.

Lion is *Panthera leo*

Tiger is *Panthera tigris*

These animals are in the same genus but are different species.

4.2 Living together

Where do organisms live?

OCR B B2a
AQA B1.11.5

Different organisms live in different environments.

- The place where an organism lives is called its **habitat.**
- All the organisms of one type living in a habitat are called a **population**.
- All the populations in a habitat are a **community**.
- An ecosystem is all the living and non-living things in a **habitat**.

Our planet has a range of different ecosystems. Some of these are natural, such as woodland and lakes. Others are artificial and have been created by people, such as fish farms, greenhouses and fields of crops. Artificial ecosystems usually have less variety of organisms living there (less biodiversity). This may be caused by the use of chemicals such as weed killers, pesticides and fertilisers.

It is possible to investigate where organisms live by using various devices.

The more samples that you take in an area then the more accurate the estimate of the whole area will be.

A **quadrat** is a small square that is put on the ground to take a sample of a large area. The number of organisms in the quadrat can be counted and the size of the population in the whole area can then be estimated.

Quadrats are often used to study plants but devices such as pooters, nets and pit-fall traps can be used to sample animal populations.

Competition and adaptation

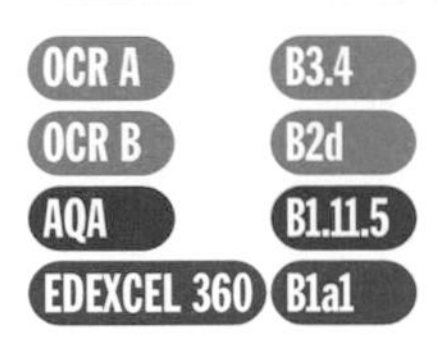

There are many different types of organisms living together in a habitat and many of them are after the same things.

This struggle for resources is called competition.

Organisms of the same species are more likely to compete with each other because they have similar needs.

The more similar the organisms, the greater the competition.
Plants usually compete for:

- **light** for photosynthesis
- **water**
- **minerals**.

Animals usually compete for:

- **food** to eat
- **water** to drink
- **mates** to reproduce with
- **shelter**.

Because there is constant competition between organisms, the best suited to living in the habitat survive. Over many generations the organisms have become suited to their environment.

The features that make organisms well suited to their environment are called adaptations.

These adaptations have all been brought about by natural selection.

Habitats such as the Arctic and deserts are difficult places to live because of the extreme conditions found there. Animals and plants have special adaptations so they can survive:

Polar bears have:	Cacti have:	Camels have:
a large body that holds heat	leaves that are just spines to reduce surface area	a hump that stores food as fat
thick insulating fur	deep or widespread roots	thick fur on top of the body for shade
a thick layer of fat under the skin	water stored in the stem	thin fur on the rest of the body
white fur that is a poor radiator of heat and provides camouflage		

Flowering plants also show adaptations. Some are adapted to being **pollinated** by insects and some are pollinated by wind.

Wind pollination is a more random process but it does not 'cost' the plant in making nectar.

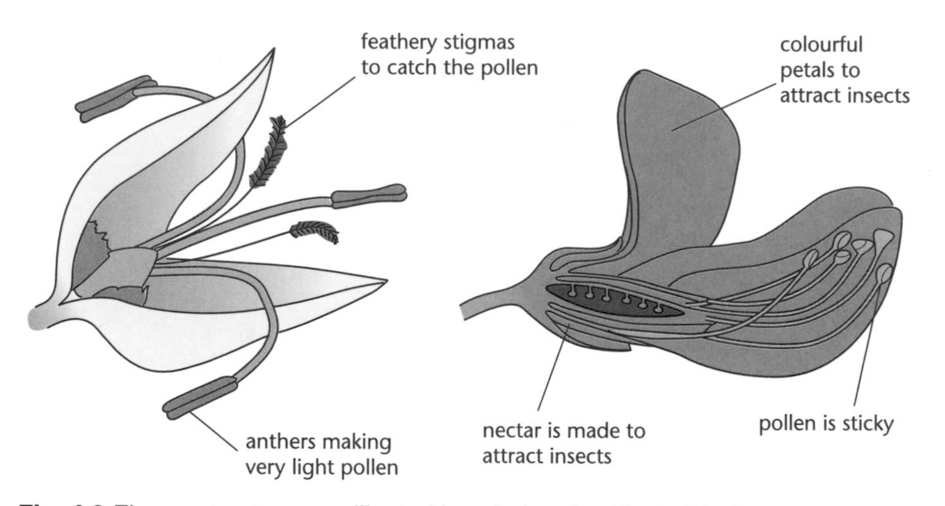

Fig. 4.3 Flower structure – pollinated by wind and pollinated by insects.

Relationships

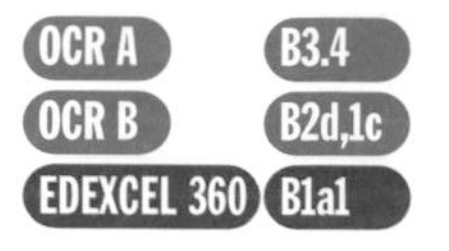

Predator and prey

Organisms form different types of relationships with other organisms in their habitat.

One of the most common is that of predator and prey.

A predator hunts and kills another animal for food. The animal that is eaten is called the prey.

Sometimes two different *y* axes are given on these graphs because the numbers of predators and prey may be very different.

The numbers of predators and prey in a habitat will vary and will affect each other. The size of the two populations can be plotted on a graph called a **predator–prey graph**.

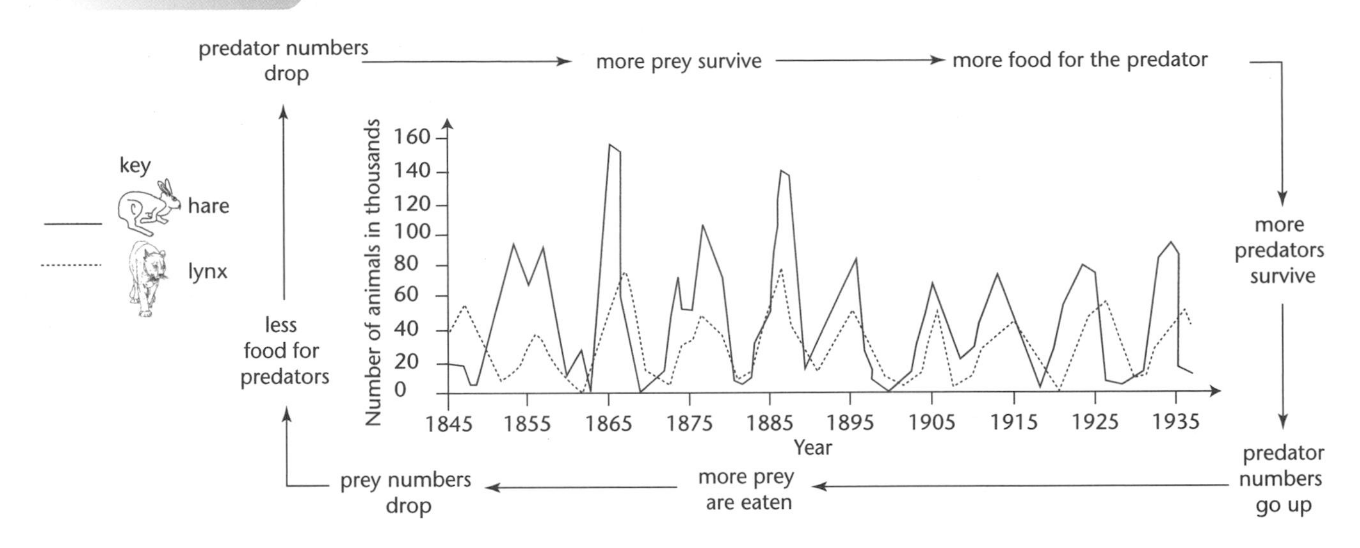

Fig. 4.4 A predator–prey graph for lynx and hares.

Parasite and host

Sometimes one organism may not kill another organism but may take food from it while it is alive.

A parasite lives on or in another living organism called the host, causing it harm.

A well-adapted parasite will not kill its host because then it will need to find another one.

Many diseases such as **malaria** are caused by parasites feeding on a host. The parasite in malaria is a single-celled organism that feeds on humans, who are the host. The organism is injected into the blood stream by a mosquito. This is also acting as a parasite but it known as a **vector** for malaria because it spreads the disease-causing organism without being affected by it.

Mutualism

Instead of trying to eat each other, some different types of organisms work together.

When two organisms of different species work together so that both gain, this is called mutualism.

What do the oxpeckers and the buffalo gain in this relationship?

Examples of this type of relationship are:

- Oxpeckers and buffalos: the oxpecker birds eat the parasites on the backs of the buffalo.
- Nitrogen-fixing bacteria in the roots of pea plants: the bacteria give the plants nitrates and they gain sugars from the plant.

Food production

Plants make their own food by a process called **photosynthesis**. They take in **carbon dioxide** and **water** and turn it into **sugars**, releasing **oxygen** as a waste product. The process needs the **energy** from sunlight and this is trapped by the green pigment **chlorophyll**.

Look back at the equation for respiration on page 11. It is the reverse of this equation.

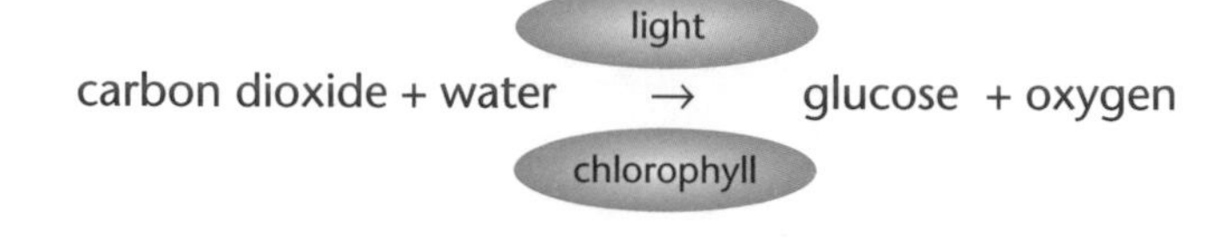

$$6CO_2 + 6H_2O \rightarrow C_6H_{12}O_6 + 6O_2$$

Once plants have made sugars such as glucose they can convert the glucose into many different things:

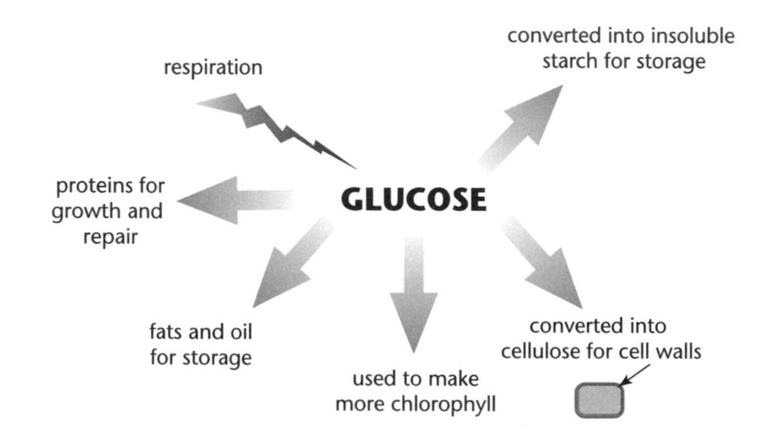

Limiting factors

The rate of photosynthesis can be increased by providing:

- more **light**
- more **carbon dioxide**
- an **optimum temperature**.

Any of these factors can be limiting factors.

> **KEY POINT**
> **A limiting factor is something that controls how fast a reaction will occur.**

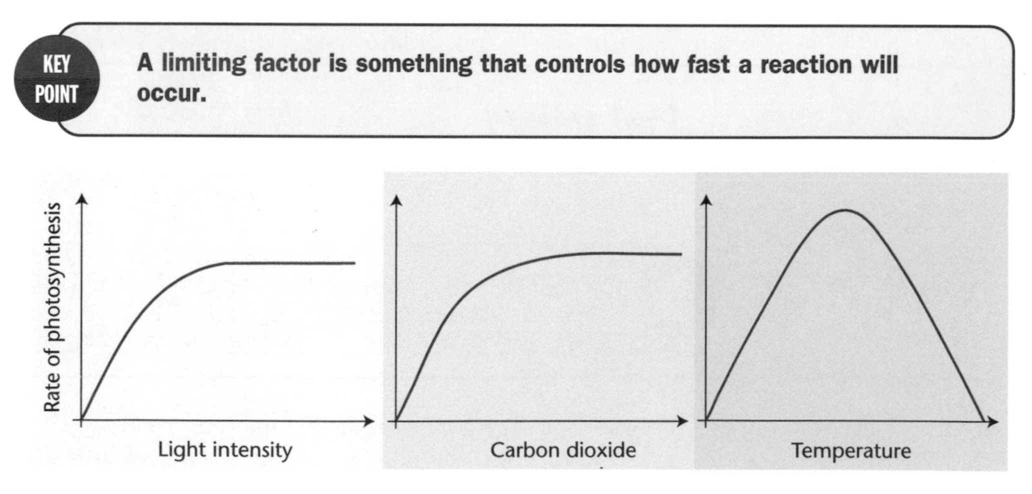

Fig. 4.5 Limiting factors for photosynthesis.

If more light is provided, it will increase photosynthesis because more energy is available. After a certain point something else will limit the rate.

More carbon dioxide will again increase the rate up to a point because more raw materials are present. Increasing temperature will make enzymes work faster but high temperatures prevent enzymes from working.

Respiration and photosynthesis

Many people think that plants respire at night and photosynthesise during the day.

In fact, plants carry out respiration all the time. Fortunately for us, during the day they photosynthesise much faster than they respire. This means that they make enough oxygen and food for us.

Energy transfer

EDEXCEL 360 B1a1

A **food chain** shows how food passes through a community of organisms. It enters the food chain as **sunlight** and is trapped by the **producers**. These are the **green plants**. The energy then passes from organism to organism as they eat each other.

Often the waste from one food chain can be used by decomposers to start another chain.

The mass of all the organisms at each step of the food chain can be measured. This can be used to draw a diagram that is similar to a **pyramid of numbers**. The difference is that the area of each box represents the mass of all the organisms, not the number.

This type of diagram is called a pyramid of biomass.

The reason that a pyramid of biomass is shaped like a pyramid is that energy is lost from the food chain in different ways as the food is passed along.

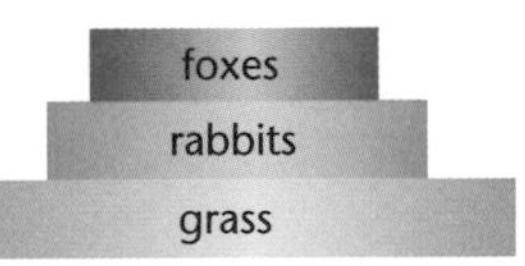

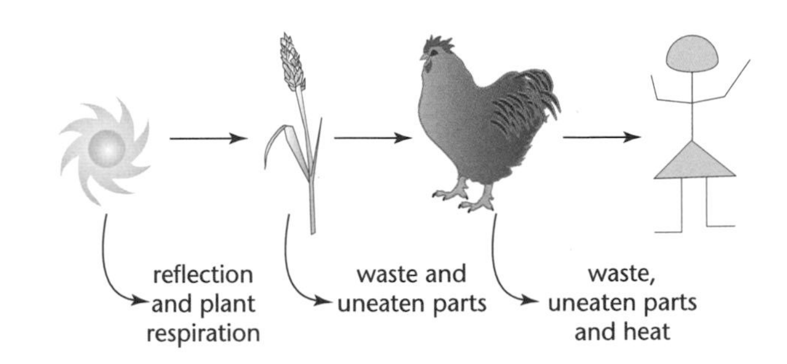

Fig. 4.6 Where energy is wasted in the food chain.

This means that less energy is lost if a person eats plant products than if they eat meat. This is because the food only goes through one transfer rather than two.

4.3 Human impact on the environment

Growth in populations

OCR B B2g
AQA B.1.11.8
EDEXCEL 360 B1a1

The human population on Earth has been increasing for a long time but it is now going up more rapidly than ever. The rate of increase is increasing and this is called **exponential growth**.

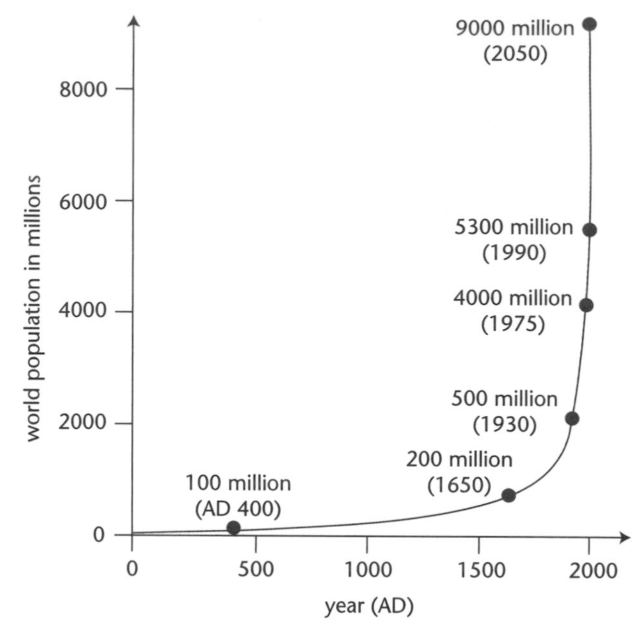

Fig. 4.7 Population growth.

This increase in the population is having a number of effects on the environment:

an increasing population

- more raw materials are being used up such as fossil fuels and minerals
- more land is being taken up to be used for building, quarrying, farming and dumping waste
- more waste is being produced which can lead to pollution

Pollution

Modern methods of food production and the increasing demand for energy have caused many different types of **pollution**.

Pollution is the release of substances into the environment that harm organisms.

The table shows some of the main polluting substances that are being released into the environment.

Polluting substance	Main source	Effects on the environment
carbon dioxide	burning fossil fuels	greenhouse effect
CFCs	fridges and aerosols	destroy the ozone layer
fertilisers	intensive farming	pollute rivers and lakes
herbicides	intensive farming	some cause mutations
methane	cattle and rice fields	greenhouse effect
sewage	human and farm waste	pollutes rivers and lakes
sulfur dioxide	burning fossil fuels	acid rain

Some scientists disagree over how severe the greenhouse effect is and what is actually causing it. **See page 68.**

The **greenhouse effect** is caused by a build-up of certain gases, such as carbon dioxide and methane, in the atmosphere. These gases trap the heat rays as they are radiated from the Earth. This causes the Earth to warm up. This is similar to what happens in a greenhouse. This could lead to changes in the Earth's climate and a large rise in sea level.

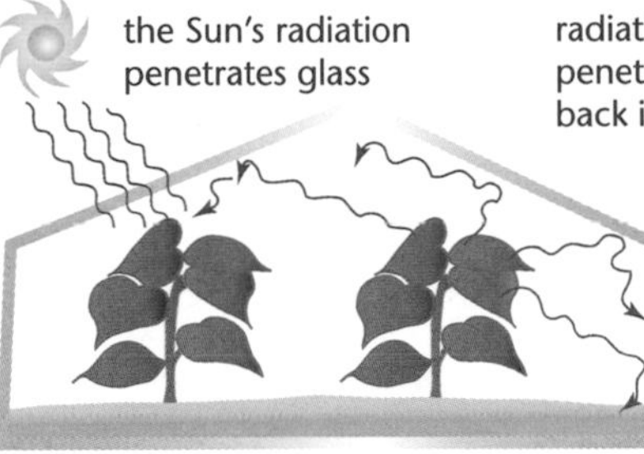

Fig. 4.8 The greenhouse effect.

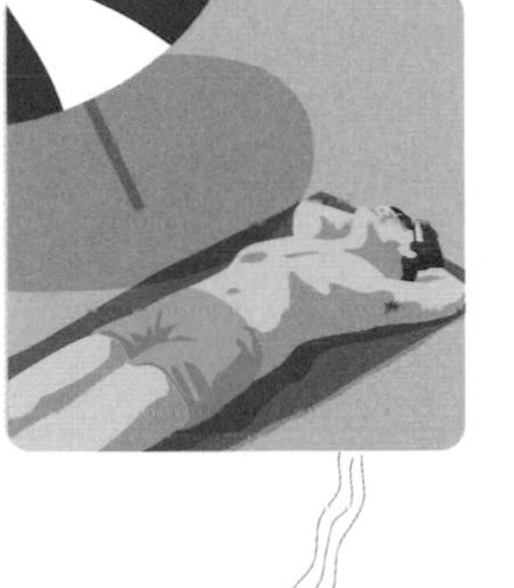

Acid rain is caused by the burning of fossil fuels that contain some **sulfur** (can also be spelt 'sulphur') impurities. This gives off sulfur dioxide, which dissolves in rainwater to form **sulfuric acid**. This falls as acid rain.

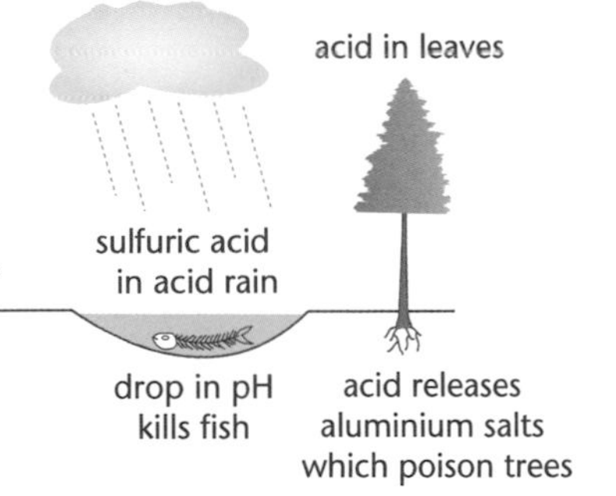

Fig. 4.9 Acid rain.

Ozone depletion is caused by the release of chemicals such as **CFCs** which come from the breakdown of refrigerators and aerosol sprays. **Ozone** helps protect us from harmful UV radiation and so depletion may lead to more skin cancer.

Pollution indicators

Some organisms are more sensitive to pollution than others. If we look for these organisms, it can tell us how polluted an area is.

On land, lichens grow on trees and stone.

Some lichens are killed by lower levels of pollution than other types.

In water some animals, such as the rat-tailed maggot, can live in polluted water but others, such as mayfly larvae, live only in clean water.

Fig. 4.10 Lichens can tell us how polluted the environment is.

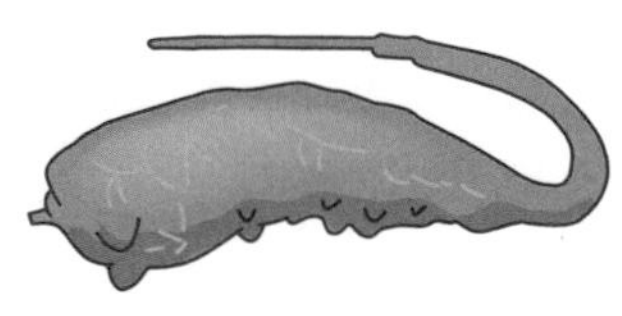

Fig. 4.11 Rat-tailed maggot.

Over-exploitation

OCR A B3.4
OCR B B2g
AQA B1.11.8
EDEXCEL 360 B1a1

Deforestation

As well as causing pollution, the increasing demands for food, land and timber have caused people to cut down large areas of forests.

Cutting down large areas of trees is called deforestation.

Deforestation has led to:

- Less carbon dioxide being removed from the air by the trees and carbon dioxide being released when the wood is burnt.
- The destruction of habitats that contain some rare species.

Overhunting and extinction

Some animals have been hunted until their numbers have been dramatically reduced.

Fig. 4.12 Extinct and endangered animals.

Many species of whales have been hunted for food, oil and other substances. Their numbers now are very low and people are trying to protect them.

Other organisms have not been so lucky. Their numbers have decreased so far that they have completely died out. This is called **extinction**. Organisms do become extinct naturally, but people have often increased the rate either directly or indirectly by:

- changing the climate
- destroying habitats
- pollution
- competition
- over-hunting.

Other organisms are at risk of becoming extinct and are **endangered**.

4.4 Conservation

Biodiversity

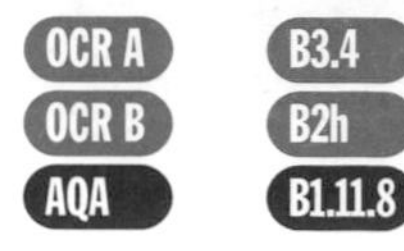

B3.4
B2h
B1.11.8

Many people believe that is wrong for humans to damage natural habitats and cause the death of animals and plants. They believe that it is important to keep a wide variety of different animals and plants alive.

The variety of different organisms that are living is called biodiversity.

It is natural for organisms to become extinct but in many cases humans are speeding up this process.

There are many reasons given for trying to maintain biodiversity:

- Losing organisms may have unexpected effects on the environment, such as the erosion caused by deforestation.
- Losing organisms may have effects on other organisms in their food web.
- Some organisms may prove to be useful in the future, for breeding, producing drugs or for their genes.

Looking after the environment

OCR B B2h

To be able to save habitats and organisms, people have set up many different schemes.

The attempt to preserve habitats and keep species alive is called conservation.

Many zoos are now more involved in conservation rather than just showing animals to people.

There are a number of different ways that conservation programmes can work:

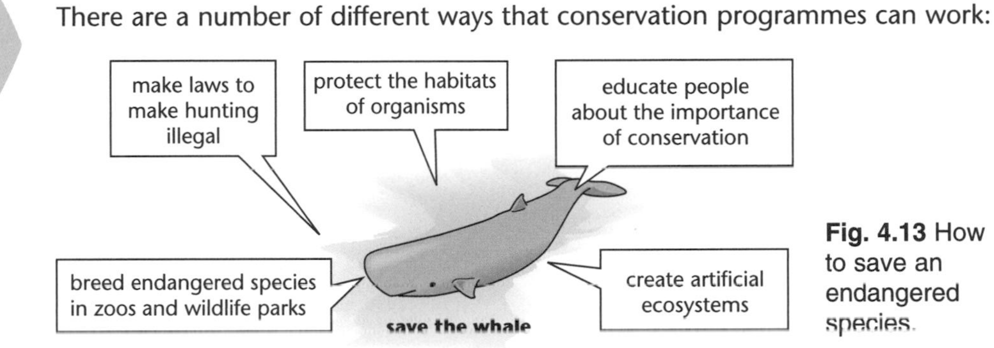

Fig. 4.13 How to save an endangered species.

Sustainable development

B3.4
B2h
B1.11.8

If the human population is going to continue to increase, it is important that we meet the demand for food and energy without causing pollution or over-exploitation.

Providing for the increasing population without using up resources or causing pollution is called sustainable development.

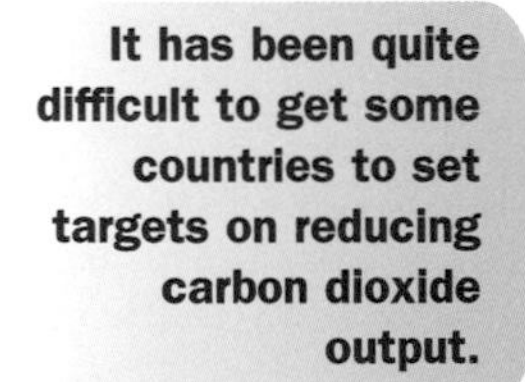

To make sure that development is sustainable a lot of planning is needed at local, national and international levels.

In 1992, over 150 nations attended a meeting in Brazil called the Earth Summit. They agreed on ways in which countries could work together to achieve sustainable development.

They agreed to:

- reduce pollution from chemicals such as carbon dioxide. This can be done by cutting down on the waste of energy or by using sources of energy that do not produce carbon dioxide
- reduce hunting of certain animals, such as whales.

The document that they signed was called Agenda 21.
In 2002, a World Summit on Sustainable Development was held in Johannesburg, South Africa, to monitor progress.

HOW SCIENCE WORKS

OCR B B2g
AQA B1.11.8
EDEXCEL B1a1

Global warming – fact or theory?

The idea that the Earth's climate has changed over long periods in the past is a well-accepted **fact** among most people, but what may have caused this?

The theory that changes in our climate could be caused by changes in the carbon dioxide levels in the atmosphere was first put forward by a Swedish scientist called Svante Arrhenius in 1896.

We now call this theory the **greenhouse effect** or **global warming**.

But what is the evidence that global warming can be caused by increasing carbon dioxide levels?

Scientists have used several sets of **data** to try and provide evidence for this theory. One important set comes from the Antarctic Vostok ice core.

Vostok is a remote place in Antarctica where a hole has been drilled down into the ice over 3.3 kilometres deep. The ice core that has been removed contains bubbles of air that were trapped at different times many thousands of years ago. The gas has been analysed to tell us the levels of carbon dioxide and the temperature of the air in the past. Graphs have been plotted that look like this:

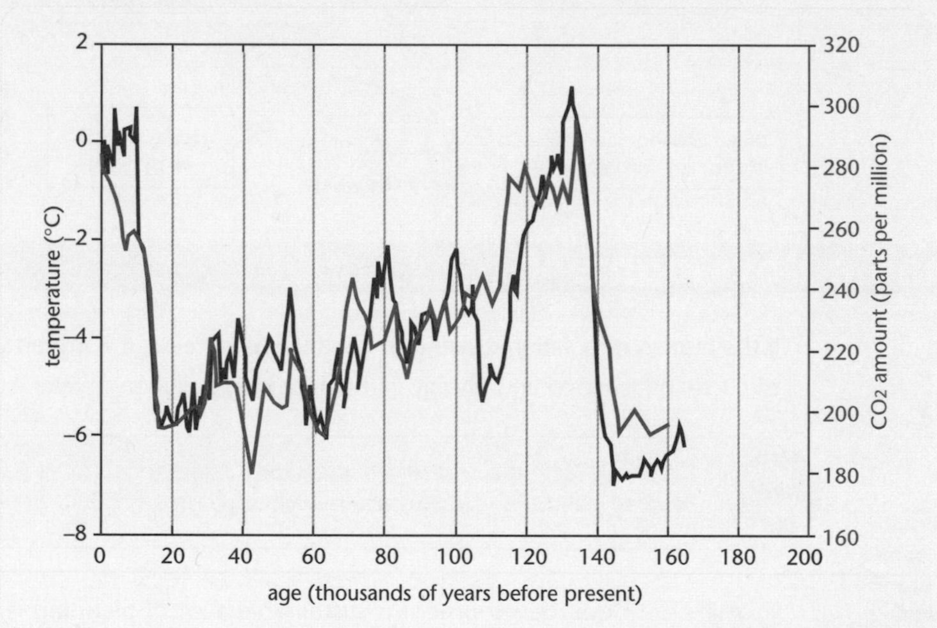

Looking at the graph, the levels of carbon dioxide and the temperature seem to follow a very similar pattern. Scientists say that there is a **correlation** between carbon dioxide levels and the temperature. This provides evidence for the theory.

However, this does not mean that one factor was **caused** by the other.

Even if one factor was caused by the other, it is possible to put forward two different theories:

- As the carbon dioxide levels increased, the Sun's rays were trapped and this caused the Earth to warm up.
- When the Earth warmed up, more animals survived and gave off more carbon dioxide.

Most scientists believe the first theory is correct, but theories are only ways of explaining data. They are not proven facts.

HOW SCIENCE WORKS

OCR B B2h

Making decisions about conservation is not always easy

Protecting whales has often been one of the main aims of conservation groups.

Records of whale hunting go back to 6000 BC. Whales have been hunted for their meat and for oil. The main countries involved in hunting have been Japan, Iceland and the United States.

In the 1800s, more modern methods of whaling reduced whale numbers dramatically and a number of whale species became endangered. Now an international group called the International Whaling Commission (IWC) exists to control whale hunting. Since 1985, commercial whaling has been banned, but some people object to this.

Norway has been hunting minke whales since 1993, killing about 600 every year.

Japan still hunts whales for 'scientific research'. According to official IWC figures, in the 2004–2005 whaling season, 601 minke whales were caught in coastal regions of Japan. Three sperm whales and 51 Bryde's whales were also taken.

In 2005, the research programme increased the quota of minke whales to 900 and, more controversially, added fin whales to the programme. This move has sparked a great deal of controversy among anti-whaling nations, in particular because fin whales are listed as endangered under the Convention on International Trade on Endangered Species. Starting in 2007, Japan plans to harvest up to 50 humpback whales and 50 fin whales annually.

Making decisions about protecting endangered species is not always very easy and clear cut. An example of this is the attempt to protect the grey whale.

A tribe of North American natives, the Makah, have hunted whales for thousands of years. They rely on the whales for food and the hunting is part of their culture. When whale numbers dropped, they had to give up hunting, and their culture and identity suffered. They now want to start hunting again on a small scale but the IWC has banned all hunting. Should they be allowed to hunt?

The arguments are still continuing.

- Should the Makah be allowed to kill a small number of whales a year to keep their culture alive?
- May this lead to other groups wanting to start hunting whales again?

HOW SCIENCE WORKS *Questions*

1. Why is it difficult to be sure about the size of whale populations? [2]
2. Discuss arguments for and against the Japanese approach to whale hunting. [2]
3. Why do some people think that the Makah tribe should be allowed to hunt a small number of whales? [2]

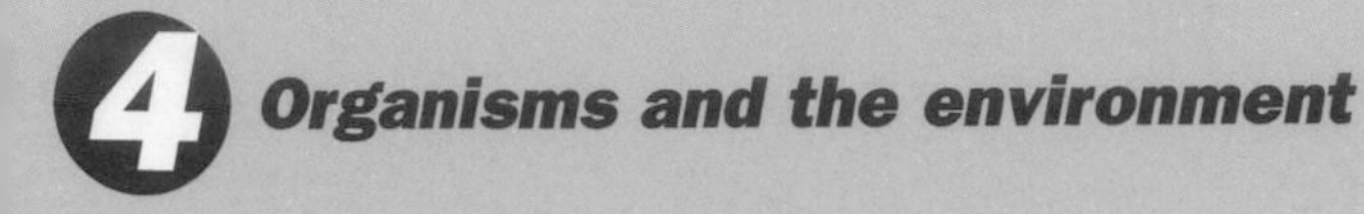

Exam practice questions

1. Which of the following is **not** an effect of acid rain?
   - **A** lowers the pH of lakes
   - **B** poisons trees with aluminium salts
   - **C** increases the temperature of the atmosphere
   - **D** kills fish **[1]**

2. Certain fungi grow on the roots of pine trees. The fungi take food from the tree roots. The fungi absorb minerals from the soil and some pass into the roots. The fungi are acting as:
   - **A** parasites
   - **B** mutualistic partners
   - **C** predators
   - **D** prey **[1]**

3. A large number of greenfly live on an oak tree. This collection of greenfly is described as:
   - **A** a community
   - **B** an ecosystem
   - **C** a class
   - **D** a population **[1]**

4. The boxes contain some types of animal and some descriptions.
   Draw straight lines to join each **type of animal** to the correct **description**.

Type of animal	Description
amphibian	has wet scales
fish	moist permeable skin
invertebrate	an animal without a backbone
mammal	covered in hair or fur
reptile	has dry scales

[3]

5. Organisms are adapted to the environment that they live in.
   Explain how the following characteristics help the organism survive:
   - **(a)** Camels store large amounts of fat in their humps. **[2]**
   - **(b)** Some cacti have deep roots that pass straight down whereas other types of cacti have shallow roots that spread out a long distance. **[3]**
   - **(c)** Polar bears are large animals with very small ears for the size of their body. **[2]**
   - **(d)** The larvae of many insects do not feed on the same type of food as the adult insect. **[1]**

Exam practice questions

6. A scientist wanted to estimate the number of greenfly and ladybirds in a large field.

(a) Explain how he would use a quadrat to estimate the numbers of the two animals in the field. **[3]**

(b) The scientist estimated the numbers at different times of the year and plotted the results on a graph.

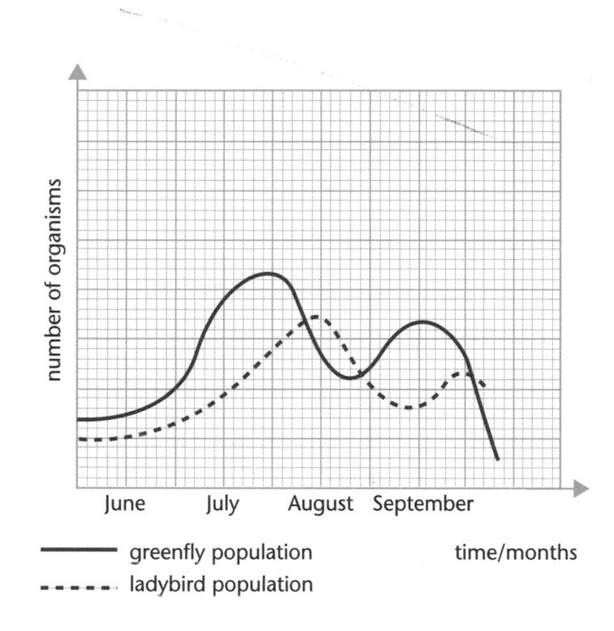

(i) Explain the shape of the scientist's graph **[3]**

(ii) What name is usually given to this type of graph? **[1]**

Chapter 5 Cells and growth

The following topics are covered in this chapter:

- *Cell structure and function*
- *Transport in cells*
- *Growth and asexual reproduction*
- *Cellular processes*

5.1 Cell structure and function

Animal and plant cells

Plant and animal cells have a number of structures in common.

They all have:

- a **nucleus** that carries genetic information and controls the cell
- a **cell membrane** which controls the movement of substances in and out of the cell
- **cytoplasm** where most of the chemical reactions happen.

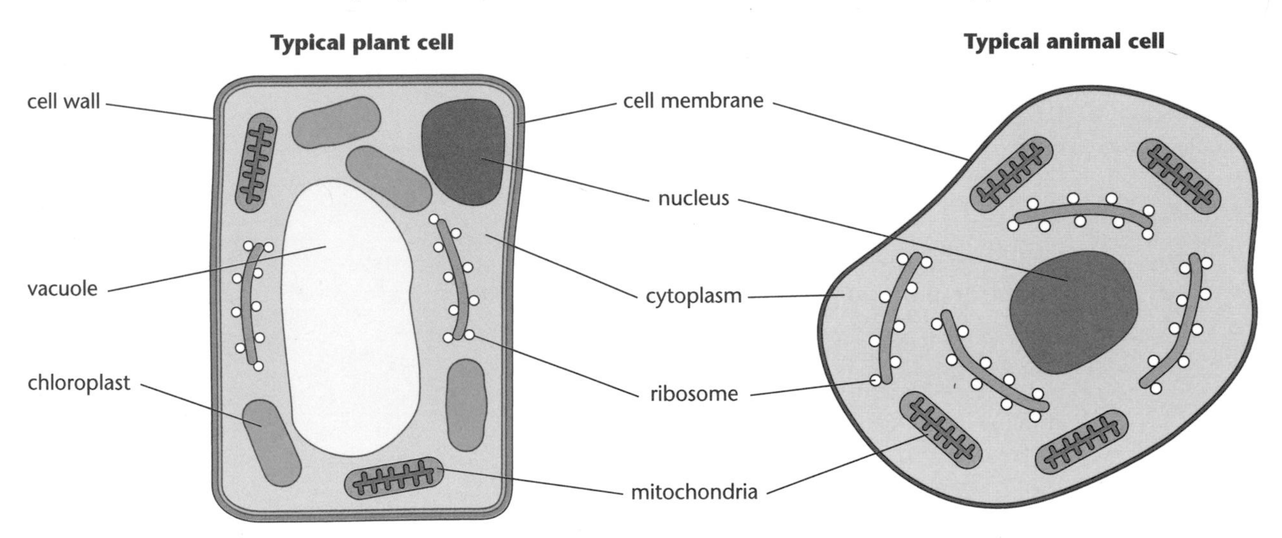

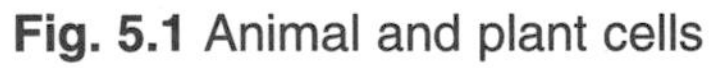
Fig. 5.1 Animal and plant cells

There are four main differences between plant and animal cells:

- plant cells have a strong **cell wall** made of cellulose, animal cells do not
- plant cells have a large permanent **vacuole** containing cell sap, vacuoles in animal cells are small and temporary
- plant cells may have **chloroplasts** containing chlorophyll for photosynthesis. Animal cells never contain chloroplasts.

Plant and animal cells also have many smaller structures in the cytoplasm. These can be seen by using an electron microscope.

Mitochondria are the site of respiration in the cell and **ribosomes** are where proteins are made.

Enzymes

OCR A B4.2
OCR B B3a
AQA B2.11.6

KEY POINT **Enzymes are biological catalysts. They are produced in all living organisms and control all the chemical reactions that occur.**

Most of the chemical reactions that occur in living organisms would happen too slowly without enzymes. Increased temperatures would speed up the reactions, but using enzymes means that the reactions are fast enough at 37 °C. These reactions include photosynthesis, respiration and protein synthesis.

Enzymes are protein molecules. They are made of a long chain of amino acids that is folded up to make a particular shape. They have a slot or a groove, called the **active site**, into which the substrate fits. The reaction then takes place and the products leave the enzyme.

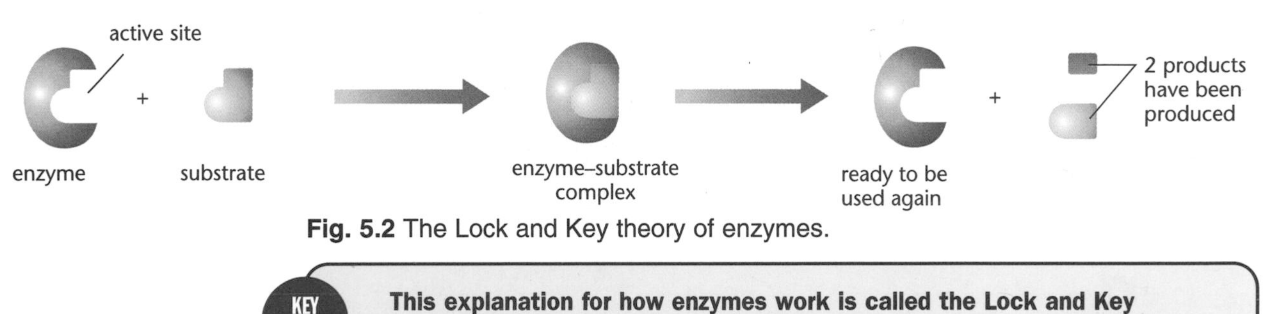

Fig. 5.2 The Lock and Key theory of enzymes.

KEY POINT **This explanation for how enzymes work is called the Lock and Key theory.**

This theory explains why an enzyme will only work on one type of substrate. They are described as **specific**.

Enzymes work best at a particular temperature and pH. This is called the **optimum** temperature or pH.

If the temperature is too low then the substrate and the enzyme molecules will not collide so often and the reaction will slow down.

If the shape of the enzyme molecule changes then the substrate will not easily fit into the active site. This means that the reaction will slow down.

Various factors may cause this to happen:

- high temperatures
- extremes of pH.

If the shape of the enzyme molecule is irreversibly changed then it is described as being **denatured**.

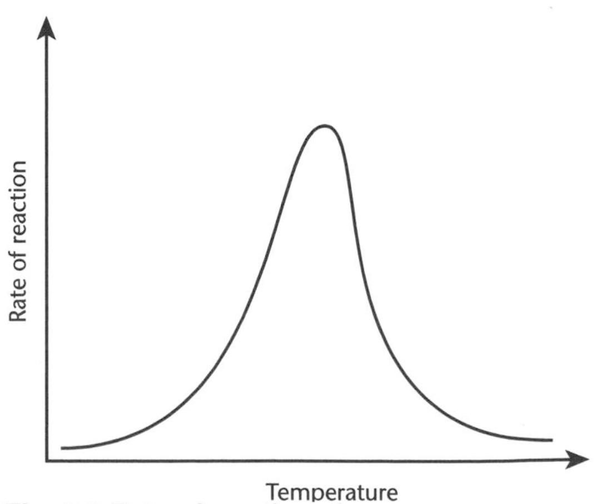

Fig. 5.3 Rate of reaction of an enzyme.

DNA and protein synthesis

OCR A	B5.2
OCR B	B3a
AQA	B11.8
EDEXCEL 360	B2.1

The nucleus controls the chemical reactions occurring in the cell. This is because it contains the genetic material. This is contained in structures called **chromosomes** which are made of a chemical called **DNA**.

DNA is a large molecule with a very important structure:

- it has two strands
- the strands are twisted to make a shape called a double helix
- each strand is a long chain of molecules called bases
- there are only four bases called A,T,G and C
- links between the bases hold the two chains together.

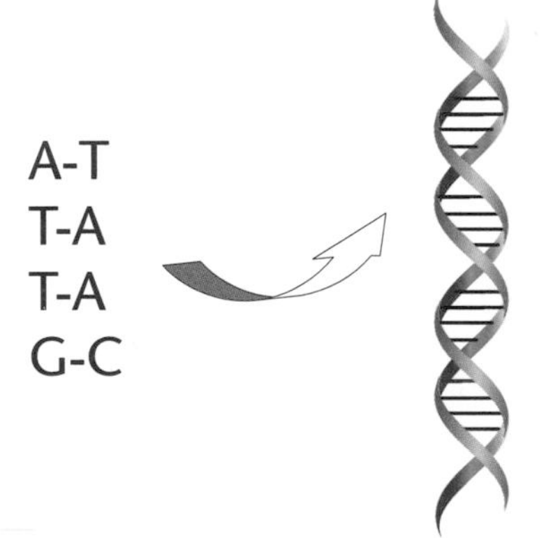

Fig. 5.4 The structure of DNA.

DNA controls the cell by carrying the code for proteins. Each different protein is made of a particular order of amino acids, so DNA must code for this order.

KEY POINT

A gene is a length of DNA that codes for the order of amino acids in one protein.

Scientists now know that each amino acid in a protein is coded for by each set of three bases along the DNA molecule.

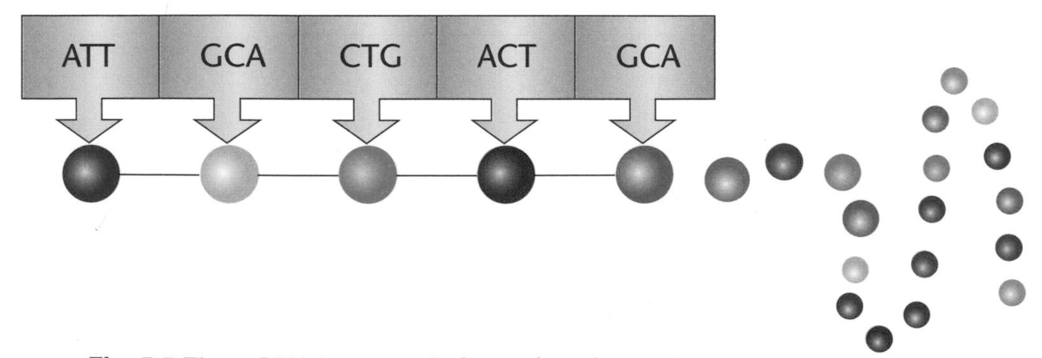

Fig. 5.5 Three DNA bases code for each amino acid.

The problem is that proteins are made on ribosomes in the cytoplasm and DNA is kept in the nucleus. The cell has to use a messenger molecule to copy the message from DNA and to carry the code to the ribosomes. This molecule is called **RNA**.

5.2 Growth and asexual reproduction

Cell division

OCR A B5.1
OCR B B3a

Before a cell divides, two things must happen. First new cell structures such as mitochondria must be made. Then the DNA must copy itself. The structure of DNA allows this to happen in a rather neat way.

- The double helix of DNA unwinds and the two strands come apart or 'unzip'.
- The bases on each strand attract their complementary bases and so two new molecules are made.

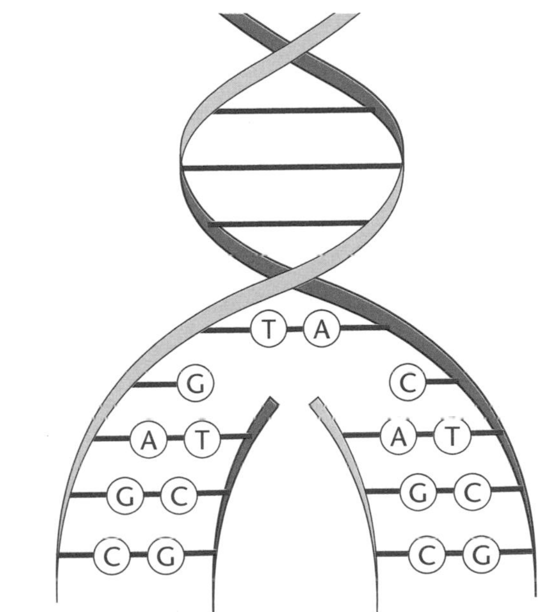

Fig. 5.6 DNA replication.

Mitosis and meiosis

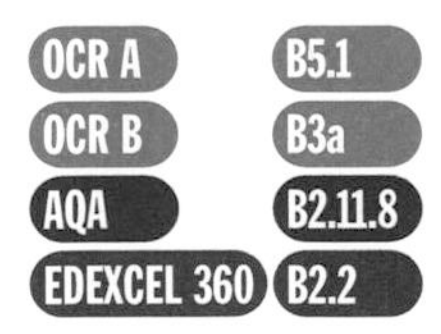

Cells divide for a number of reasons. There are two types of cell division and they are used for different reasons.

KEY POINT **There are two types of cell division, meiosis and mitosis.**

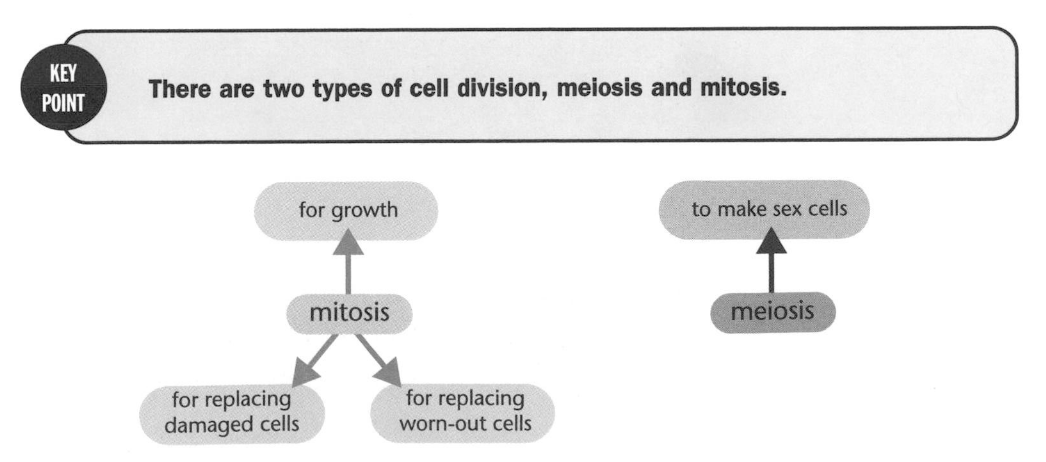

Fig. 5.7 Uses of mitosis and meiosis.

In **mitosis**, two cells are produced from one. As long as the chromosomes have been copied exactly, then each new cell will have the same number of chromosomes and the same information as each other and the parent plant.

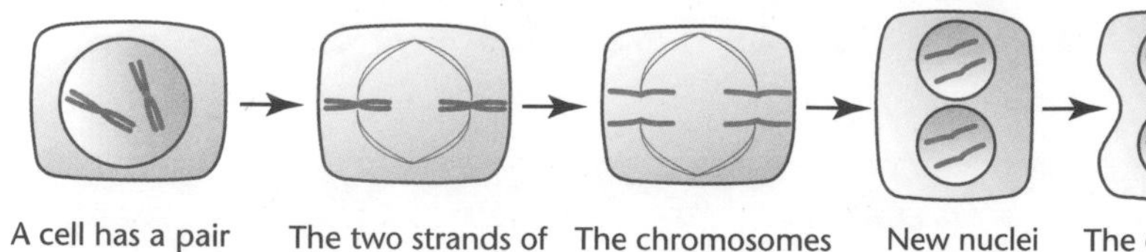
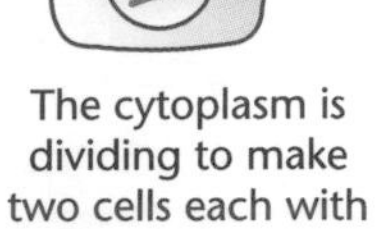

Fig. 5.8 Mitosis.

In **meiosis**, the chromosomes are also copied once but the cell divides twice. This makes four cells each with half the number of chromosomes, one from each pair.

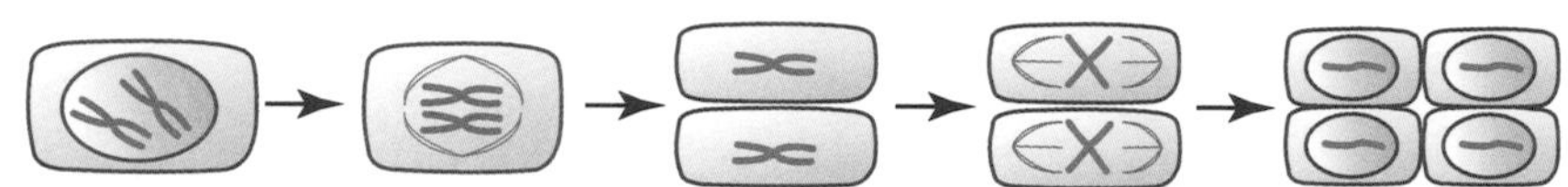

Fig. 5.9 Meiosis.

Cells with one chromosome from each pair are called haploid and can be used as gametes. When two gametes join, the diploid or full number of chromosomes is produced.

Growth and development

OCR A B5.3
OCR B B3d, 3e
AQA B2.11.8
EDEXCEL 360 B2.2

Most organisms are multi-cellular. This means that they are made up of a large number of cells. This has a number of advantages:

- it allows organisms to be larger
- it allows the cells to become specialised for different jobs.

When gametes join at fertilisation, this produces a single cell called a **zygote**. This soon starts to divide many times by mitosis to produce many identical cells.

These cells then start to become specialised for different jobs.

The production of different types of cells for different jobs is called differentiation.

These differentiated cells can then form tissues and organs. Once a cell has differentiated, it cannot form other types of cells. Although it has the same genes as all the other cells, many are turned off so it only makes the proteins it needs.

Some cells in the embryo and in the adult keep the ability to form other types of cells. They are called **stem cells**. Scientists are now trying to use stem cells to replace cells that have stopped working or been damaged. This may have the potential to cure a number of diseases.

Embryonic stem cells can produce a greater range of different cells than adult stem cells.

Human growth and development

OCR B B3e

Humans grow at different rates at different times of their lives.

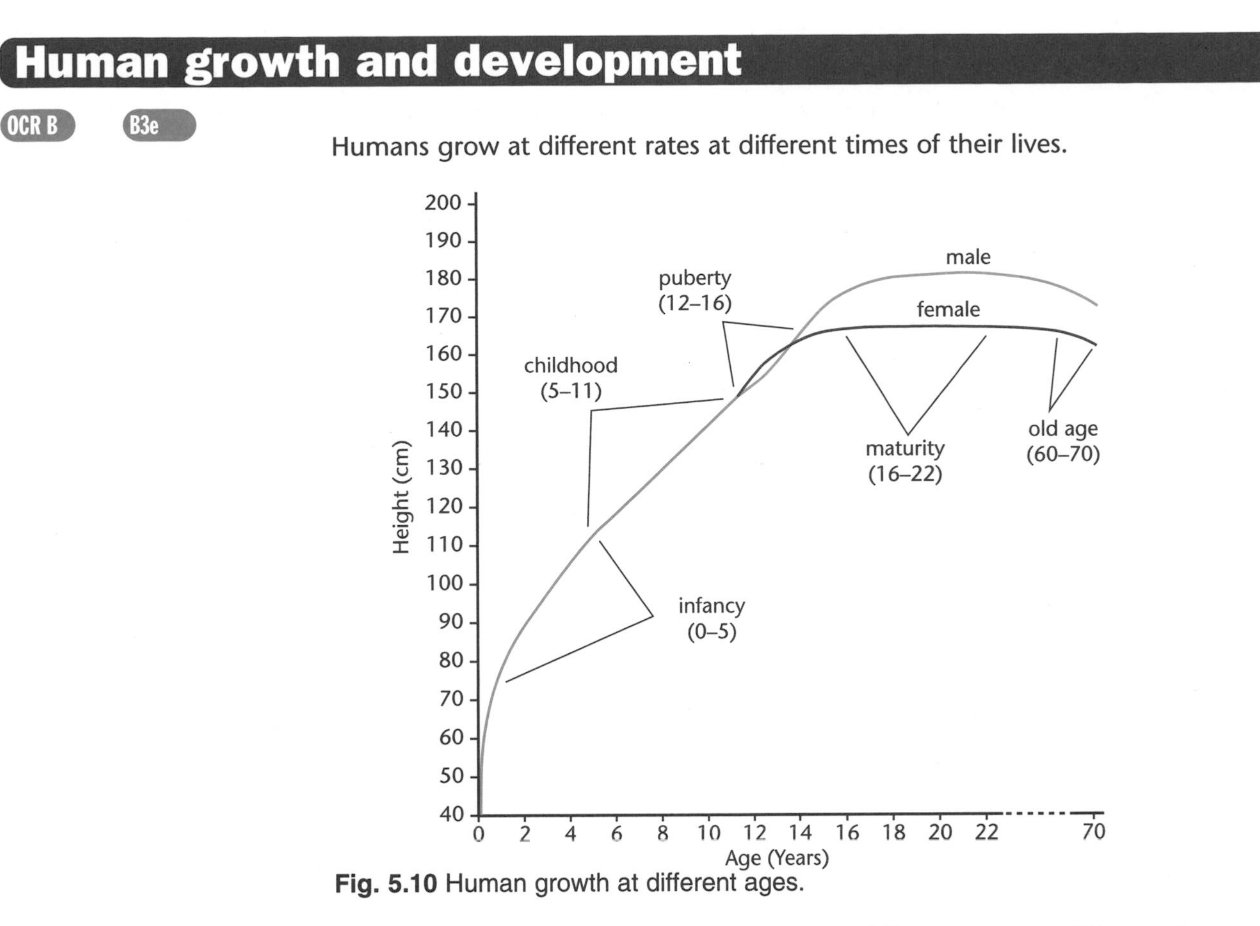

Fig. 5.10 Human growth at different ages.

The various parts of their body also grow at different rates at different times.

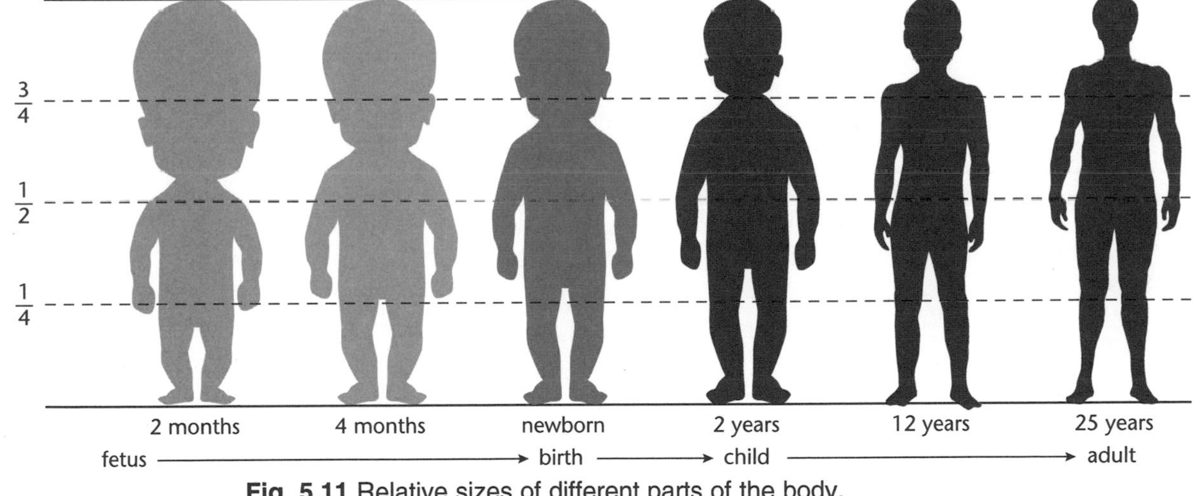

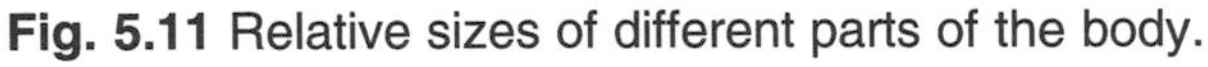

Fig. 5.11 Relative sizes of different parts of the body.

Plant growth and development

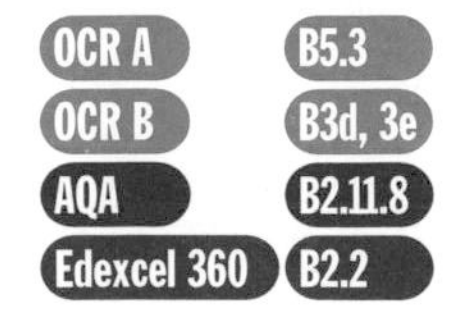

Like animals, plants grow by making new cells through mitosis. The cells then differentiate into tissues like xylem and phloem. These tissues then form organs such as roots, leaves and flowers.

Growth in plants is different to animal growth in a number of ways:

- Plant cells enlarge much more than animal cells after they are produced. This increases the size of the plant.
- Cells tend to divide at the ends of roots and shoots. This means that plants grow from their tips.

Plant cells that can produce new types of cells are called meristematic.

- Animals usually stop growing when they reach a certain size, but plants carry on growing.
- Many plant cells keep the ability to produce new types of cells, in animals only stem cells can do this.

Growth in plants is controlled by chemicals called **plant growth substances** or **plant hormones**. They control:

- the rate and direction that roots and shoots grow
- the time that plants flower
- the ripening of fruits.

KEY POINT **The main type of plant hormones are called auxins.**

By controlling the growth of plants, auxins can allow plants to respond to changes happening around them. This means that the roots and shoots of plants can respond to gravity or light in different ways.

	Type of growth	
change	**shoots**	**roots**
gravity	away = negatively geotropic	towards = positively geotropic
light	towards = positively phototropic	away = negatively phototropic

Auxins change the direction that roots and shoots grow by changing the rate that the cells elongate.

Growing towards the light means that shoots can gather more light for photosynthesis.

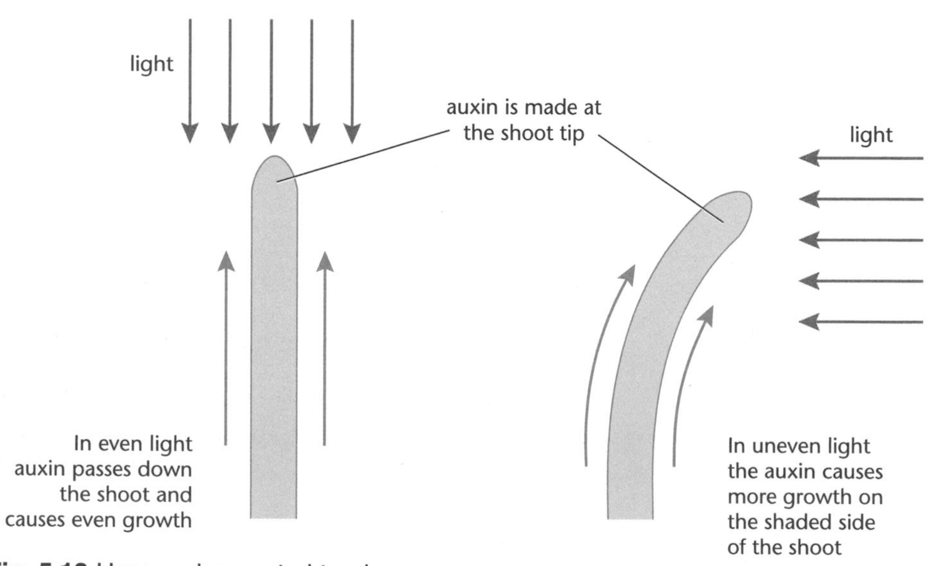

Fig. 5.12 How auxins control tropisms.

Gardeners can use auxins to help them control the growth of their plants.

kill weeds in lawns

control seed germination

control the ripening of fruit

treat cuttings to make them produce roots

Fig. 5.13 Uses of auxins.

Cloning

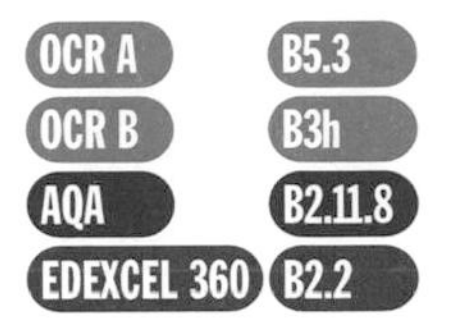

A clone is an identical genetic copy.

Because plants have many cells that have not yet differentiated, it is easy to produce identical copies of useful plants. Plants may also produce identical copies naturally. This is called asexual reproduction.

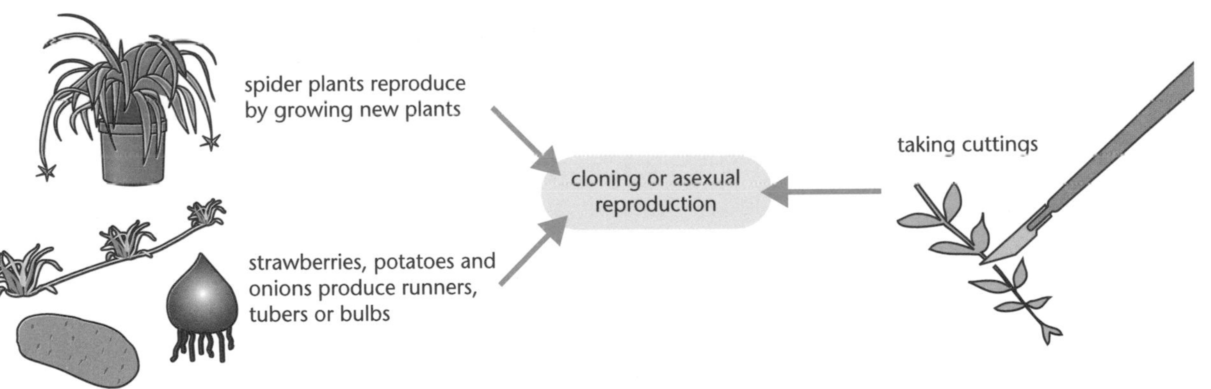

Fig. 5.14 Different types of asexual reproduction.

Producing plants by cloning has a number of advantages and disadvantages.

Advantages	Disadvantages
you know what you are going to get because all the plants will be genetically identical to each other and the parent	the population of plants will be genetically very similar – there will be little variety
you can produce many plants that do not flower very often or are difficult to grow from seeds	because the plants all have the same genes, a disease or change in the environment could kill them all

5.3 Transport in cells

Transport across the cell membrane

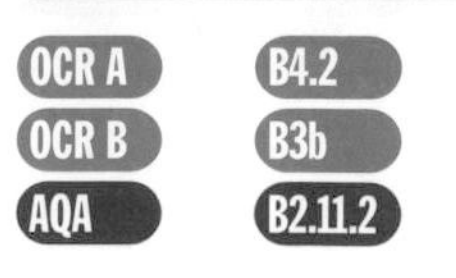

Substances can pass across the cell membrane by three different processes:

- diffusion
- osmosis
- active transport.

KEY POINT **Diffusion is the movement of a substance from an area of high concentration to an area of low concentration.**

Diffusion works because particles are always moving about in a random way.

The rate of diffusion can be increased in a number of ways:

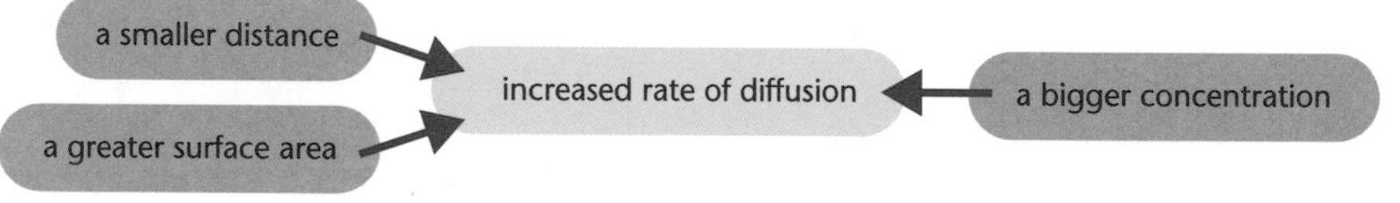

Fig. 5.15 Factors that increase diffusion rate..

Osmosis is really a special type of diffusion. It involves the diffusion of water.

Remember that an area of high water concentration is usually called a dilute solution.

Osmosis is the movement of water across a partially permeable membrane from an area of high water concentration to an area of low water concentration.

The cell membrane is an example of a partially permeable membrane. It lets small molecules through, such as water, but stops larger molecules.

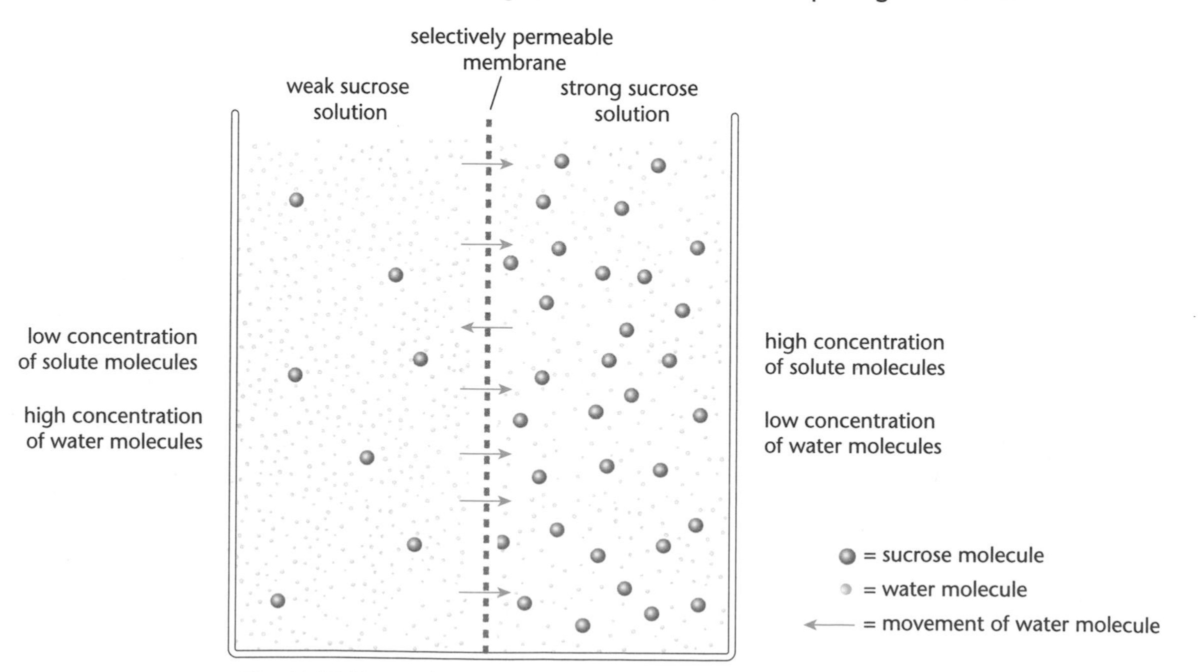

Fig. 5.16 Osmosis.

Turgid cells are very important in helping to support plants. Plants with flaccid cells often wilt.

When plant cells gain water by osmosis, they swell. The cell wall stops them from bursting. This makes the cell stiff or **turgid**. If a plant cell loses water, it goes limp or **flaccid**.

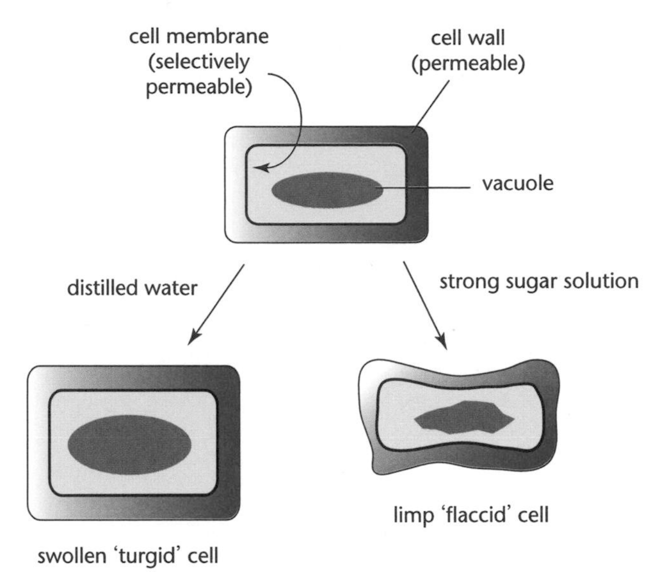

Fig. 5.17 Osmosis in plant cells.

Animal cells do not behave in the same way because they do not have a cell wall. They will either swell up and burst or shrink if they gain or lose water.

Sometimes substances have to be moved from a place where they are in low concentration to where they are in high concentration. This is in the opposite direction to diffusion and is called **active transport**.

KEY POINT

Active transport is the movement of a substance against the diffusion gradient with the use of energy from respiration.

Examples of diffusion

OCR A B4.2
OCR B B3b

There are many examples of the movement of substances by diffusion in plants and animals.

Site	Substances	Direction of movement	Special features
small intestine	digested food materials	from the small intestine into the blood	villi and microvilli give the intestine a large surface area
lungs	gases	oxygen moves from the air to the blood and carbon dioxide moves the opposite way	the alveoli have a large surface area, a moist lining, a good blood supply and a thin wall
placenta	food materials, waste products and gases	oxygen and food moves from the mother to the fetus. Carbon dioxide and waste moves the opposite way	the placenta has a large surface area, a rich blood supply and a thin wall
neurones	transmitter substances	move across synapses	allows nerve impulses to pass from neurone to neurone
leaves	water, carbon dioxide and oxygen	water and oxygen move out of the leaves and carbon dioxide moves in	leaves have a large surface area, pores called stomata and many air spaces

5.4 Cellular processes

Where does photosynthesis occur?

OCR B B4a

Photosynthesis occurs mainly in the leaves of a plant.

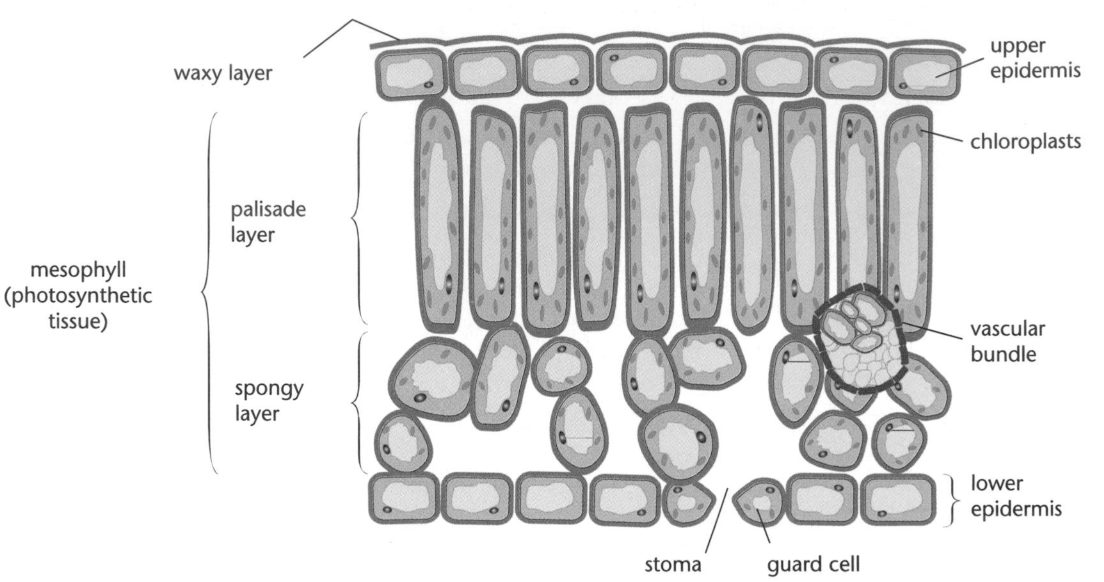

Fig. 5.18 Cross section of a leaf.

The leaves are specially adapted for photosynthesis in a number of ways.

Adaptation	How it helps photosynthesis
a broad shape	provides a large surface area to absorb light and CO_2
a flat shape	the gases do not have too far to diffuse
contains a network of veins	supplies water from the roots and takes away the products
contains many chloroplasts in the palisade layer near the top	traps the maximum amount of light
pores called stomata and air spaces	allow gases to diffuse into the leaf and reach the cells

Aerobic respiration

AQA B2.11.6
EDEXCEL 360 B2.1

The reactions of aerobic respiration take place in mitochondria.

The energy that is released by respiration can be used for many processes:
- to make large molecules from smaller ones (e.g. proteins from amino acids)
- to contract the muscles
- for mammals and birds to keep a constant temperature
- for active transport.

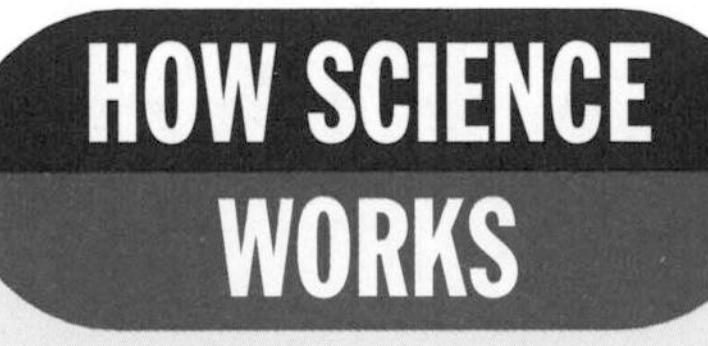

OCR A	B5.3
OCR B	B3e
AQA	B2.11.8
EDEXCEL 360	B2.2

Stem cells help dogs with dystrophy

Scientists think that a recent experiment with stem cells using dogs is an important step towards treating people.

The experiment focused on Duchenne muscular dystrophy, a muscle-wasting genetic disorder that occurs in about 1 in every 3,500 boys that are born.

Children with the disorder have trouble walking and nearly all of them lose their ability to walk between ages 7 and 12. They usually die in their 20s because of weakness in their heart and lung muscles. There is no known cure.

The scientists worked with dogs that suffer a type of dystrophy that is very much like the human one. They gave the dogs repeated injections of a kind of stem cell extracted from blood vessels.

Two dogs that were severely disabled by the disease were able to walk faster and even jump after the treatments.

This is just one example of how stem cells may be used to treat people.

One of the main problems is where to get stem cells from.

The most useful stem cells are from early embryos. These cells are called embryonic stem cells and can develop into any other type of cell. If they are taken from a cloned embryo of a person, there is also the big advantage that they will not be rejected by that person. The problem is that extracting the cells involves destroying the embryo.

The search is now on to find other sources of stem cells. There are a number of possibilities.

Setting up a stem cell bank

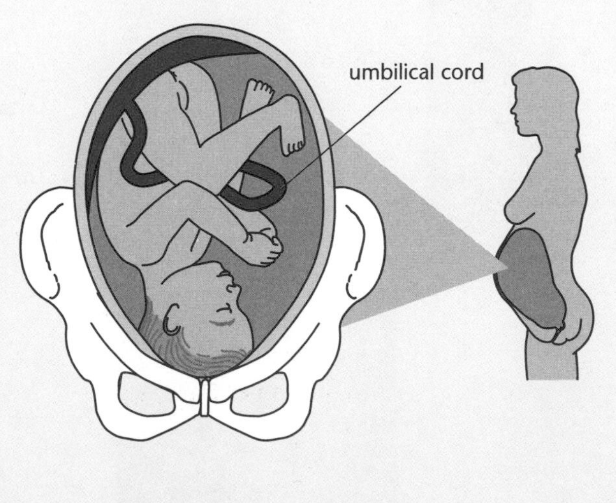

A company is setting up a new service for parents. The company will take blood from a baby's umbilical cord when it is born. The blood is then frozen and stored.

This is done because the blood contains stem cells. It is stored in the hope it could one day be used to treat certain blood diseases. The cost is £1500.

HOW SCIENCE WORKS

The stem cells found in the umbilical cord are not embryonic stem cells and can only form certain cells in the blood and immune system. The advantage is that they are much easier to obtain than adult stem cells. They are also much less likely to be rejected if given to the person who they were taken from.

They could be used to treat a number of diseases such as leukaemia. The Leukaemia Research charity say that cord blood transplants would offer the best chance of survival to about 40 children each year in Britain.

Stem cells from 'virgin birth'

Some animals, like water fleas, can produce eggs that develop into babies without being fertilised by a male. This is an example of asexual reproduction. This means that all the genetic material comes from the mother. No mammals have been found that can do this.

Scientists have now been able to get a mouse egg to start to divide without being fertilised. The ball of cells cannot produce a mouse because it always dies at an early stage.

Scientists think that they may be able to repeat this with human eggs. This means that they may be able to extract embryonic stem cells from them before they die.

Here are some people's views about these uses of stem cells. Explain why they might have these views. **[6]**

"I think that it is wrong to use cloned embryos to extract stem cells."

"I am not happy with companies charging to store blood from umbilical cords."

"I think that more people would agree to use the ball of cells produced from an unfertilised egg rather than from a cloned embryo."

Exam practice questions

1. The following structures are found in plant and animal cells.
Match words **A**, **B**, **C**, and **D**, with the numbers 1–4 in the sentences.

A mitochondria
B cell wall
C vacuole
D cell membrane

All organisms release energy from food, this largely happens in the ____1____. Cells take up water by osmosis because the ____2____ is selectively permeable. The ____3____ stores some sugars and salts. Plant cells are limited to how much water they can take up because the ____4____ resists the uptake of too much water. **[4]**

2. Complete the table putting a tick (✓) or a cross (✗) in the blank boxes.

	Osmosis	**Diffusion**	**Active transport**
Can cause a substance to enter a cell		✓	
Needs energy from respiration	✗		
Can move a substance against a concentration gradient	✗		
Is responsible for oxygen moving into the red blood cells in the lungs			✗

[4]

3. The boxes contain some chemicals that are found in the cell and some functions.
Draw straight lines to join each chemical to its correct function.

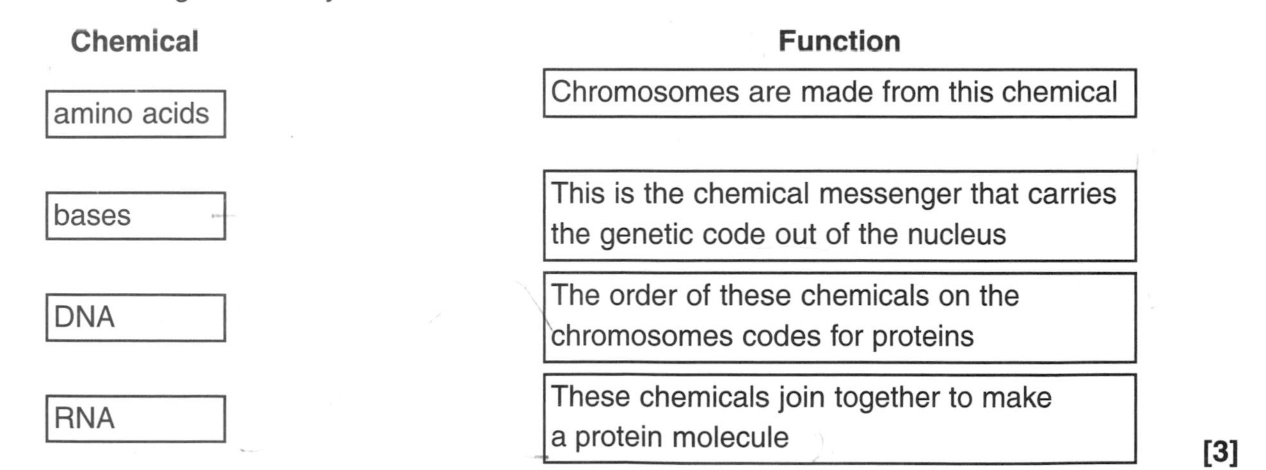

[3]

4. Complete the table by writing the correct numbers in each box.

The number of chromosomes in a human body cell.	
The number of cells made from one cell when it divides by meiosis.	
The number of chromosomes in a human sperm cell.	
The number of strands in each DNA molecule.	

[4]

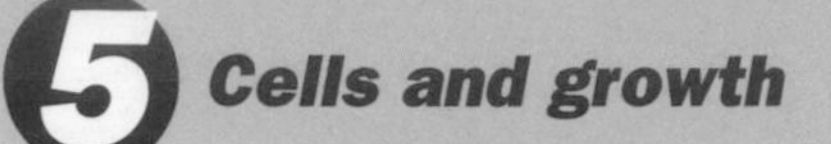

Exam practice questions

5. All the cells in the human body have about 20 000 genes.
Scientists have studied some organs to see how many of these genes are used by cells in each organ.
This number is shown on the diagram.

heart 1195
liver 2091
pancreas 1094
kidney 712
small intestine 297

(a) Write down precisely where in a cell the genes are found. **[2]**

(b) Genes code for proteins.
Explain how each gene can code for a different protein. **[2]**

(c) What percentage of its genes does each pancreas cell actually use? **[1]**

(d) Which organ on the diagram would you expect to carry out the most chemical reactions?
Explain your answer. **[2]**

6. Arthur wants to measure how fast a plant photosynthesises at different light intensities.
The diagram shows the apparatus he uses.
Arthur includes the following steps in his method:

lamp
water with sodium hydrogen carbonate
beaker
pondweed
ruler

- He uses the same piece of pondweed for the complete investigation.
- He adds sodium hydrogen carbonate to the water to provide the plant with carbon dioxide.
- He times five minutes using the minute hand of his watch and counts the number of bubbles given off.
- He counts the bubbles three different times at each light intensity.
- He repeats this with the light at different distances from the pondweed.

(a) Why does Arthur choose pondweed for his experiment? **[2]**

(b) What is the main gas found in the bubbles? **[1]**

(c) Write down the step that:

(i) helps to make Arthur's experiment a **fair test** **[1]**

(ii) makes Arthur's experiment more **reliable**. **[1]**

(d) Suggest one way that Arthur could make his experiment more **accurate**. **[1]**

Chapter 6

Ecology

The following topics are covered in this chapter:

- *Food chains and energy flow*
- *Minerals and cycles*

6.1 Food chains and energy flow

Energy flow

OCR B B4e
AQA B2.11.4

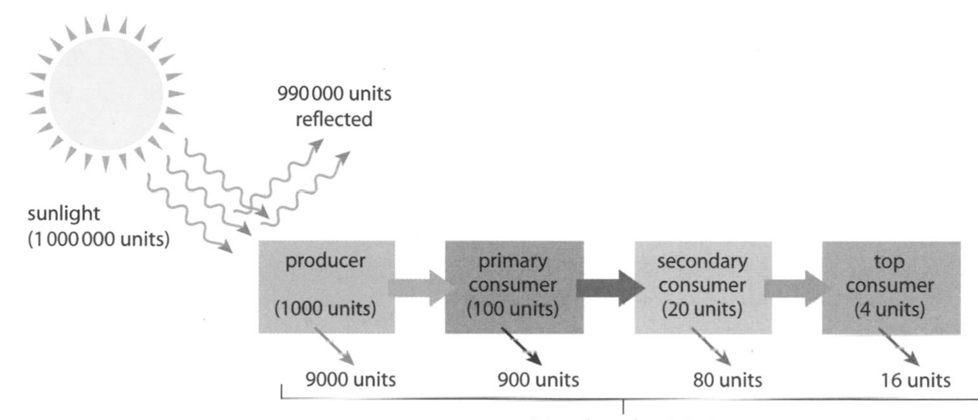

Fig. 6.1 Flow of energy through a food chain.

Remember excretion is the removal of waste products made by the body, egestion is food material that passes straight through.

Fig. 6.1 shows that biomass and energy are lost from the food chain in a number of ways:

- In waste from the organisms. This is from excretion and egestion.
- As heat when organisms respire. Birds and mammals that keep a constant body temperature will often lose large amounts of energy in this way.

This loss of energy also explains why food chains usually only have four or five steps. By then there is so little energy left, animals would not be able to find enough food.

Intensive food production

The human population is increasing and so there is a greater demand for food. This means that many farmers now use **intensive farming** methods.

Intensive farming means trying to obtain as much food as possible from the land, animals and plants.

Farmers use a number of intensive farming techniques to help increase their yield.

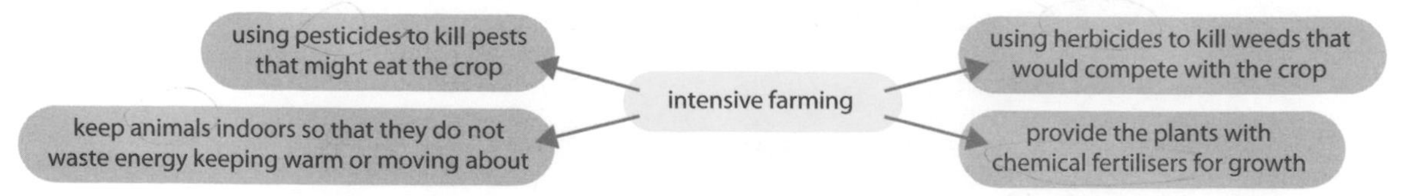

There are a number of different food production systems that use intensive methods:

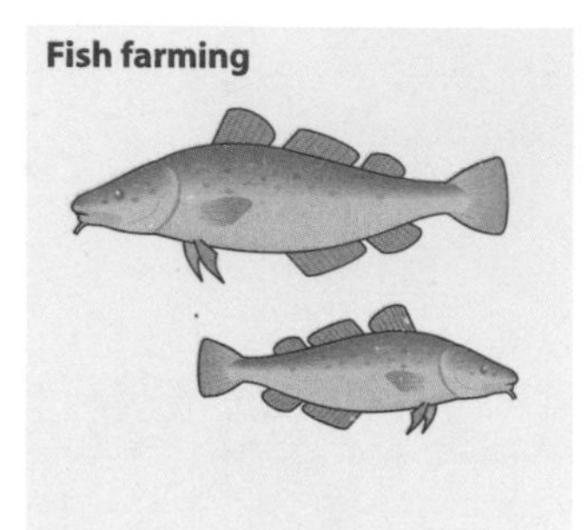

Fish are kept in enclosures away from predators. Their food supply and pests are controlled.

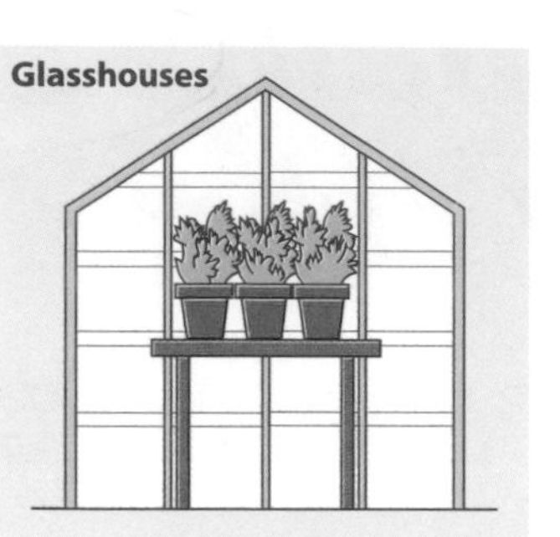

Plants can be grown in areas where the climate would not be suitable. They can also produce crops at different times of the year.

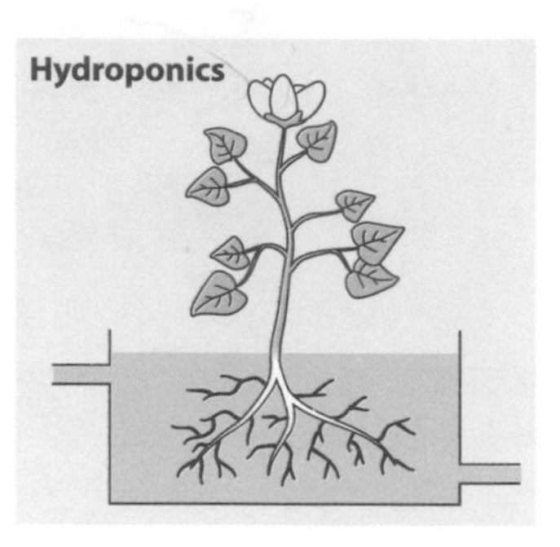

Plants are grown without soil. They need extra support but their mineral supply and pests are controlled.

Fig. 6.2 Food production systems that use intensive methods.

Organic food production

OCR B B4f

Many people think that intensive farming is harmful to the environment and cruel to animals. Farming that does not use the intensive methods is called **organic farming**.

Organic farming uses a number of different techniques:

Technique	Details
use of manure and compost	these provide minerals for the plant instead of using chemical fertilisers
crop rotation	farmers do not grow the same crop in a field year after year. This stops the build up of pests
use of nitrogen fixing crops	these crops contain bacteria that add minerals to the soil
weeding	this means that chemical herbicides are not needed
varying planting times	this can help to avoid times that pests are active
using biological control	farmers can use living organisms to help to control pests. They may eat them or cause disease

6.2 Minerals and cycles

Minerals and plants

OCR B B4d
AQA B2.11.3
EDEXCEL 360 B2.3

For healthy growth, plants have to turn the glucose made in photosynthesis into many other chemicals. Some of these are shown on page 20.

To produce these chemicals, plants need various minerals from the soil:

- nitrates as a supply of nitrogen to make amino acids and proteins
- phosphates to supply phosphorus to make DNA and cell membranes
- potassium to help enzymes in respiration and photosynthesis
- magnesium to make chlorophyll.

Without these minerals, plants do not grow properly.

The minerals are taken up by the roots by active transport. This needs energy from respiration but it means that they can be taken up even if the concentration in the soil is very low.

no nitrates
stunted growth, yellow older leaves

no phosphates
purple, younger leaves

no potassium
yellow leaves with dead spots

no magnesium
stunted growth, pale yellow leaves

Fig. 6.3 Lack of certain minerals is shown on a plant's leaves.

Decay

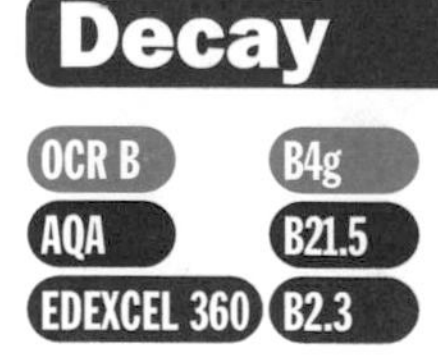

Some animals and plants die before they are eaten. They also produce large amounts of waste products. This waste material must be broken down or **decayed** because it contains useful minerals. If this did not happen, organisms would run out of minerals.

Organisms that break down dead organic material are called decomposers.

The main organisms that act as decomposers are bacteria and fungi. They release enzymes on to the dead material to digest the large molecules. They then take up the soluble chemicals that are produced. The bacteria and fungi use the chemicals in respiration and for raw materials. This type of nutrition is called **saprophytic** and decomposers are sometimes called saprophytes.

For decomposers to decay dead material they need certain conditions:

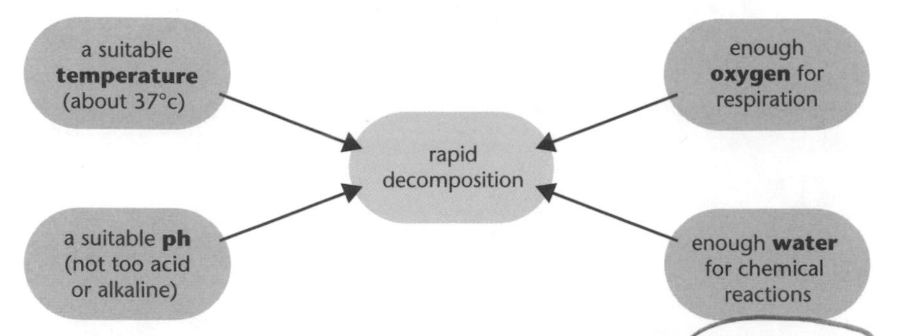

Some animals such as earthworms and woodlice are called **detritivores**. They feed on dead material and break it up into smaller pieces. This produces a larger surface area so decomposers can produce faster decay.

Fig. 6.4 An earthworm and woodlouse.

Gardeners try to produce ideal conditions for decay in their compost heaps.

Although gardeners want decay to happen in compost heaps, people do not want their food to decay before they can eat it.

KEY POINT **Food preservation methods reduce the rate of decay of foods.**

There are many ways to preserve food. Most stop decay by taking away one of the factors that decomposers need.

Preservation method	How it is done	How does it work?
canning	food is heated in a can and the can is sealed	the high temperature kills the microorganisms and oxygen cannot get into the can after it is sealed
cooling	food is kept in a refrigerator at about 5 °C	the growth and respiration of the decomposers slow down at low temperature
freezing	food is kept in a freezer at about −18 °C	the decomposers cannot respire or reproduce
drying	dry air is passed over the food	microorganisms cannot respire or reproduce without water
adding salt or sugar	food is soaked in a sugar solution or packed in salt	the sugar or salt draws water out of the decomposers
adding vinegar	the food is soaked in vinegar	the vinegar is too acidic for the decomposers

Nutrient cycles

OCR B B4h
AQA B2.11.4
EDEXCEL 360 B2.3

It is possible to follow the way in which each mineral element passes through living organisms and becomes available again for use. Scientists use nutrient cycles to show how these minerals are recycled in nature.

Carbon dioxide is returned to the air in a number of different ways, but the main process that removes it from the air is photosynthesis.

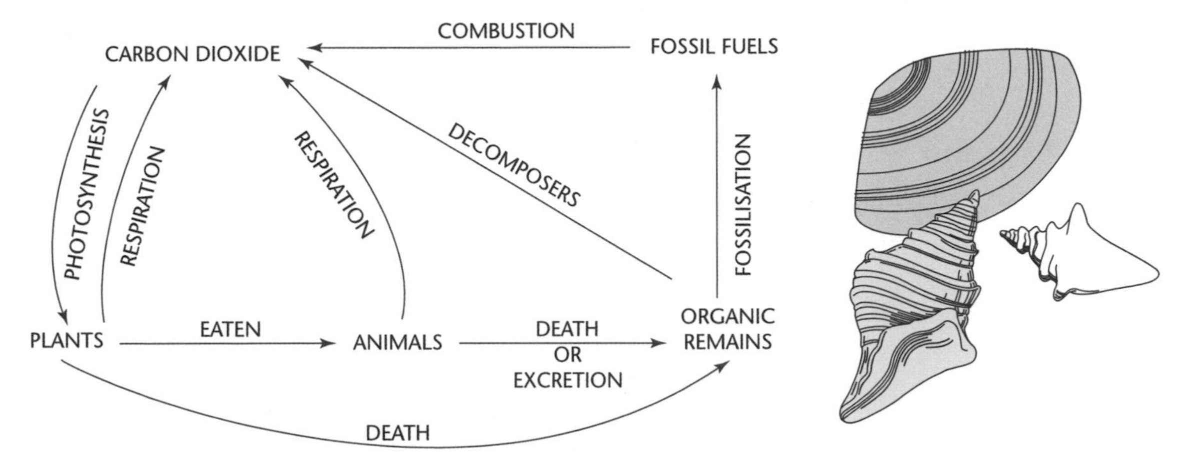

Fig. 6.5 The carbon cycle.

In the sea, carbon dioxide can also be locked up in the shells of animals. They can then form limestone. This can release carbon dioxide as it weathers or during volcanic eruption.

AQA candidates do not need to learn the nitrogen cycle.

Although the air contains about 78% nitrogen it is unreactive. It needs lightning to make it react with oxygen.

The nitrogen cycle is more complicated because, as well as the decomposers, it involves three other types of bacteria:

- **nitrifying bacteria** – these bacteria live in the soil and convert ammonium compounds to nitrates. They need oxygen to do this
- **denitrifying bacteria** – these bacteria in the soil are the enemy of farmers. They turn nitrates into nitrogen gas. They do not need oxygen
- **nitrogen fixing bacteria** – they live in the soil or in special bumps called nodules on the roots of plants from the pea and bean family. They take nitrogen gas and convert it back to useful nitrogen compounds.

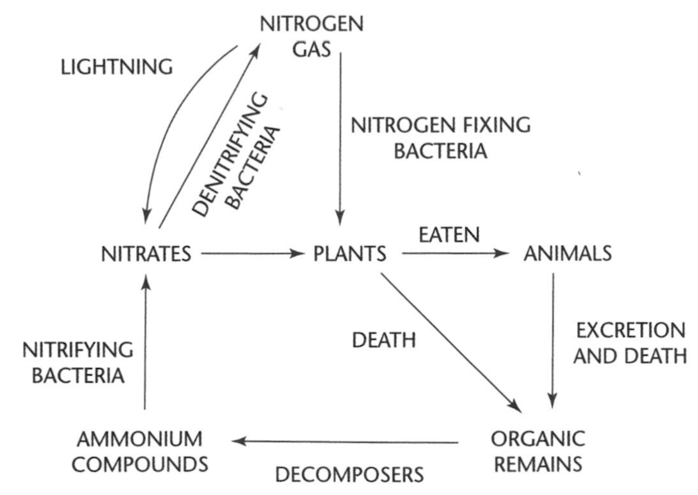

Fig. 6.6 The nitrogen cycle.

HOW SCIENCE WORKS

AQA B2.11.4
EDEXCEL 360 B2.3

Organic versus non-organic food

Organic food has now become big business. It is not necessary to go to special shops to buy organic food because all the major supermarkets have their own ranges of these foods. Sales of organic foods are increasing dramatically. Organic food sales have increased from just over £100 million in 1993/94 to £1.21 billion in 2004, more than £2 million a week.

But how do people know that food is organic?

The use of the term **organic**, when applied to food, has a legal meaning. It means the food has been grown and processed according to certain rules, known as Standards. These Standards cover every stage of organic food production, from farm to shop. These Standards usually mean that organic food is dearer than non-organic food – but is it worth paying the extra money?

Some people think so, but others do not.

These two articles give two different points of view:

“Is it doing more harm than good?”

There is no doubt that much organic food is dearer than non-organic food.

For example, an organic chicken costs about £8.50 in a supermarket, almost three times the price of a non-organic chicken.

But some scientists now think that organic food could actually be harming the environment more than food grown using pesticides and fertilisers. A recent report says that some organic farming can create greater pollution and cause more global warming.

Scientists think that some foods such as organic milk, chicken and tomatoes need more energy and land for their production than non-organic foods.

'You cannot say that all organic food is better for the environment than all non-organic food,' said one of the scientists.

'If you look carefully at the amount of energy required to produce these foods, you get a complicated picture. In some cases, the carbon footprint for organic food is larger.'

The study looked at Britain's 150 top-selling foods. It studied the energy used to grow the food, along with processing and packaging. It also looked at the by-products from the farming.

It found that organic farming can cause some environmental problems. For example, because organic chickens take longer to grow than battery hens, they had a larger effect on the environment.

HOW SCIENCE WORKS

“It must be better for us”

There is no doubt that organic food is better for people and for the environment.

Organic food is grown without the using artificial pesticides and fertilisers. It has been known for some time that pesticides and fertilisers can stay on non-organic foods. They are eaten by people and, over long periods of time, these chemicals can build up in fatty tissues. They can then become dangerous. Although there have been no long-term studies about the difference between eating non-organic versus organic foods, it is obvious that organic food is better for you.

If you consider organic meat, it contains no antibiotics or growth hormones. All organic food contains no artificial additives, preservatives, colourings or flavourings and no hydrogenated fats. These fats can cause heart disease. The artificial additives in non-organic foods can make food last beyond its natural sell-by date, and make it look more colourful. At best, these additives are unnecessary and at worst, they may lead to cancer and could be causing damage that nobody has yet discovered.

Some people say that organic food can cause food poisoning. Organic food has not been linked to any case of food poisoning in any year since records began. But perhaps best of all, organic food tastes better and contains 50% more nutrients, minerals and vitamins than produce that has been intensively farmed.

HOW SCIENCE WORKS Questions

Try thinking about the answers to these questions:

1. The first article mentions ‘scientists’ and a ‘report’? What effect do you think this has on people? [2]
2. Does the first article give a balanced argument? What does it concentrate on? [2]
3. Why do you think that people have such different views on this subject? [2]

Exam practice questions

1. The following substances are needed for plant growth.
Match words **A**, **B**, **C**, and **D**, with the numbers **1–4** in the sentences.

A auxins
B carbon dioxide
C nitrates
D magnesium

Plants need ___**1**___ as the raw material for photosynthesis.
They also need chlorophyll which contains the mineral ___**2**___
To grow plants need to convert sugars into proteins using ___**3**___
The growth of the plant is controlled by ___**4**___ **[4]**

2. The table contains some farming methods.
Put an (✓) next to any method that would **not** be used by organic farmers.

Spreading manure on the fields	
Spraying chemical pesticides	
Killing weeds using weedkillers	
Rotating their crops	

[2]

3. The boxes contain some bacteria found in the nitrogen cycle and some roles.
Draw straight lines to join each **bacteria** to its correct **role**.

Bacteria	**Role**
nitrogen fixing bacteria	breakdown organic remains
decomposing bacteria	convert ammonium compounds to nitrates
nitrifying bacteria	convert nitrate to nitrogen gas
denitrifying bacteria	convert nitrogen gas to nitrogen compounds

[3]

Exam practice questions

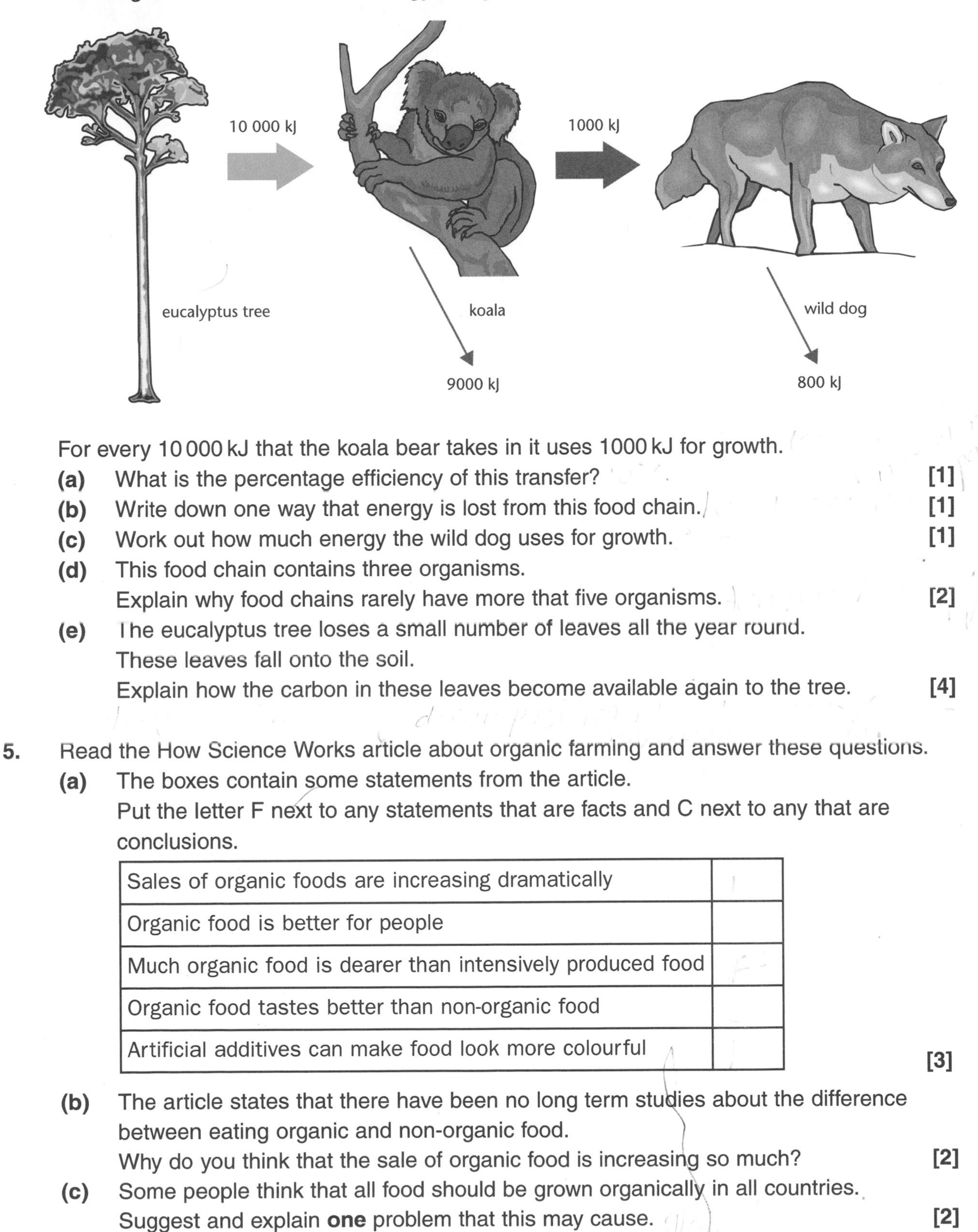

4. The diagram shows the flow of energy along a food chain.

For every 10 000 kJ that the koala bear takes in it uses 1000 kJ for growth.

(a) What is the percentage efficiency of this transfer? **[1]**

(b) Write down one way that energy is lost from this food chain. **[1]**

(c) Work out how much energy the wild dog uses for growth. **[1]**

(d) This food chain contains three organisms.
Explain why food chains rarely have more that five organisms. **[2]**

(e) The eucalyptus tree loses a small number of leaves all the year round.
These leaves fall onto the soil.
Explain how the carbon in these leaves become available again to the tree. **[4]**

5. Read the How Science Works article about organic farming and answer these questions.

(a) The boxes contain some statements from the article.
Put the letter F next to any statements that are facts and C next to any that are conclusions.

Sales of organic foods are increasing dramatically	
Organic food is better for people	
Much organic food is dearer than intensively produced food	
Organic food tastes better than non-organic food	
Artificial additives can make food look more colourful	

[3]

(b) The article states that there have been no long term studies about the difference between eating organic and non-organic food.
Why do you think that the sale of organic food is increasing so much? **[2]**

(c) Some people think that all food should be grown organically in all countries.
Suggest and explain **one** problem that this may cause. **[2]**

Chapter 7

Further physiology

The following topics are covered in this chapter:

- ***Homeostasis***
- ***Digestion***
- ***Transport in plants and animals***
- ***Nerves, synapses and the brain***

7.1 Homeostasis

Principles of homeostasis

KEY POINT **It is vital that the internal environment of the body is kept fairly constant. This is called homeostasis.**

The different factors that need to be kept constant include:

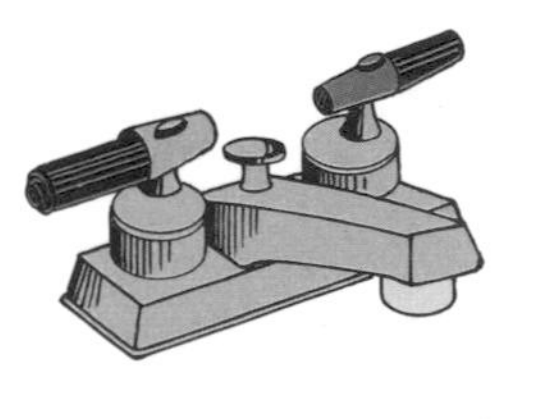

water content

temperature

sugar content

mineral content

The body has a number of automatic control systems that keep these factors at steady levels. This is important for cells to function properly.

These control systems work in the same sort of way as some artificial systems, such as the temperature control in a house.

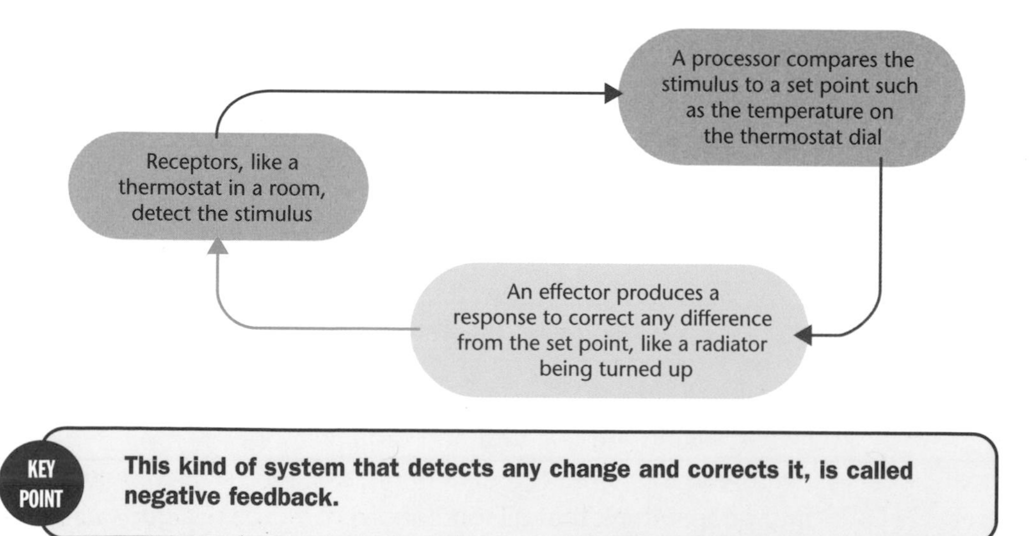

KEY POINT **This kind of system that detects any change and corrects it, is called negative feedback.**

Systems that work against each other are called antagonistic.

Some systems have more than one effector that can work in opposite directions. This means that the response can happen much faster. This is like turning the radiators up and down, and also opening or closing the windows.

Control of body temperature

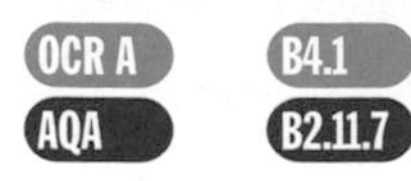

Like many factors in the body, it is important to make sure that heat gain and loss are balanced, otherwise the body temperature will change.

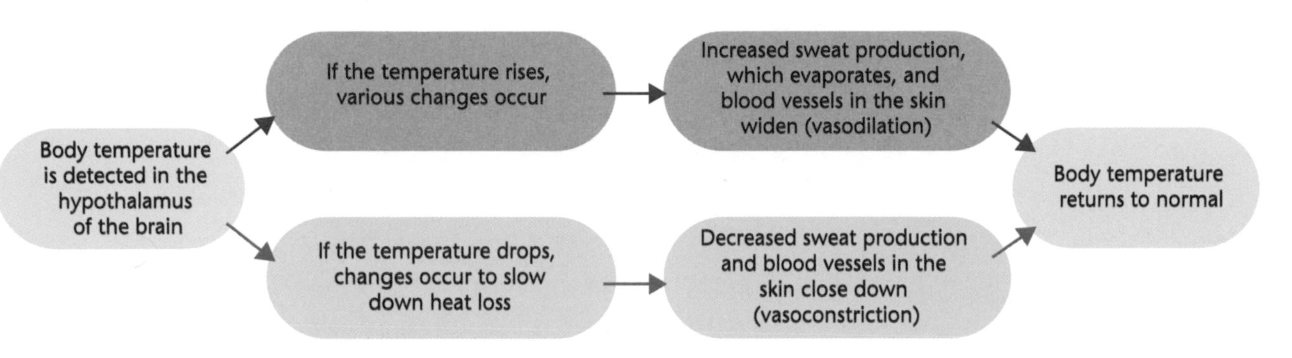

The body also has temperature receptors in the skin. These tell the brain about the external temperature so any changes in body temperature can be predicted.

Sometimes when we are ill, or in extreme conditions, temperature control can go wrong in several ways.

Heat stroke

- This is an uncontrolled increase in body temperature.
- It is caused by extreme exercise or very hot conditions.
- It can cause headaches, sickness and dizziness.
- The person should be cooled down and given liquid to stop dehydration which otherwise would stop the person sweating and increase the temperature.

Hypothermia

- This happens when the body temperature falls below 35 °C.
- The person shivers violently and may be confused.
- They should be kept warm with dry blankets.

Control of blood sugar

AQA B2.11.7

It is vital that the sugar or glucose level of the blood is kept constant. This job is performed by the **pancreas**.

Insulin is the hormone that controls the level of glucose in the blood.

When glucose levels are too high, the pancreas makes more insulin. This allows more glucose to move into the cells from the blood.

Control of water balance

OCR A B4.4

The **kidneys** remove waste such as urea from the body. They also control the water balance of the body.

The kidneys do this in the following way:

- they filter the blood to remove all small molecules
- they reabsorb useful molecules such as sugar, back into the blood
- then a certain amount of water and salts are taken back into the blood to keep their levels in balance
- the remaining waste is stored in the bladder as urine.

The amount of water that is taken back into the blood is controlled by a hormone called **ADH** which is released by the **pituitary gland**.

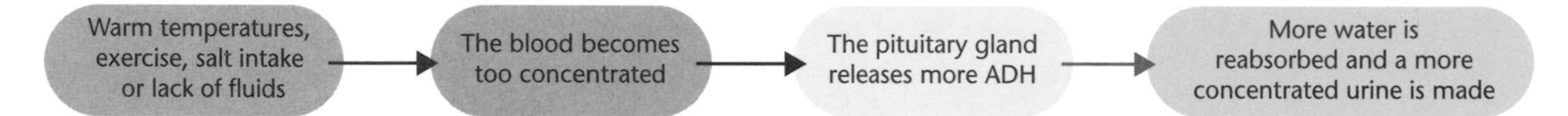

Different drugs can alter ADH release. **Alcohol** reduces ADH release and can cause too much urine to be made. **Ecstasy** can cause the opposite effect.

7.2 Transport in plants and animals

Blood and the heart

OCR B B3c

The blood is made up of a liquid called **plasma**. This carries chemicals such as hormones, antibodies and waste products around the body.

Cells are also carried in the plasma. They are adapted for different jobs.

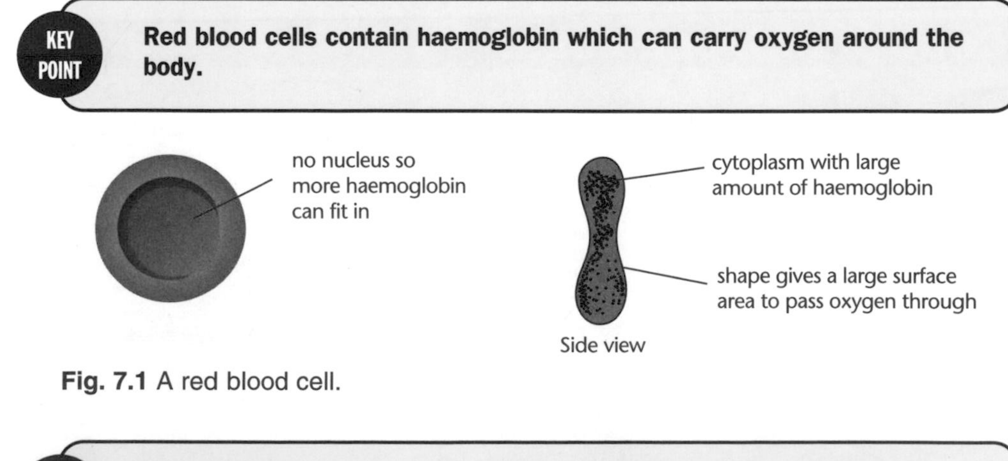

Fig. 7.1 A red blood cell.

KEY POINT **White blood cells can change shape to engulf and destroy disease organisms.**

The blood is carried around the body in arteries, veins and capillaries.

Arteries	Veins	Capillaries
carry blood away from the heart	carry blood back to the heart	join arteries to veins
have thick, muscular walls because the blood is under high pressure	have valves and a wide lumen because the blood is under low pressure	have permeable walls so that substances can pass in and out to the tissues

A double circulation keeps the pressure higher so the blood flows faster.

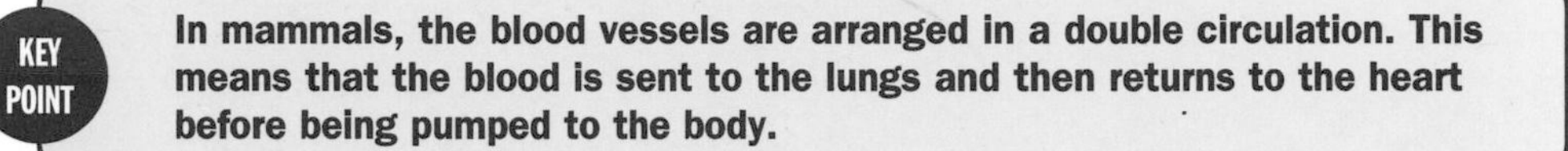

KEY POINT

In mammals, the blood vessels are arranged in a double circulation. This means that the blood is sent to the lungs and then returns to the heart before being pumped to the body.

This means that the heart has to deal with oxygenated and deoxygenated blood at the same time. The two types of blood are kept separate in different sides of the heart.

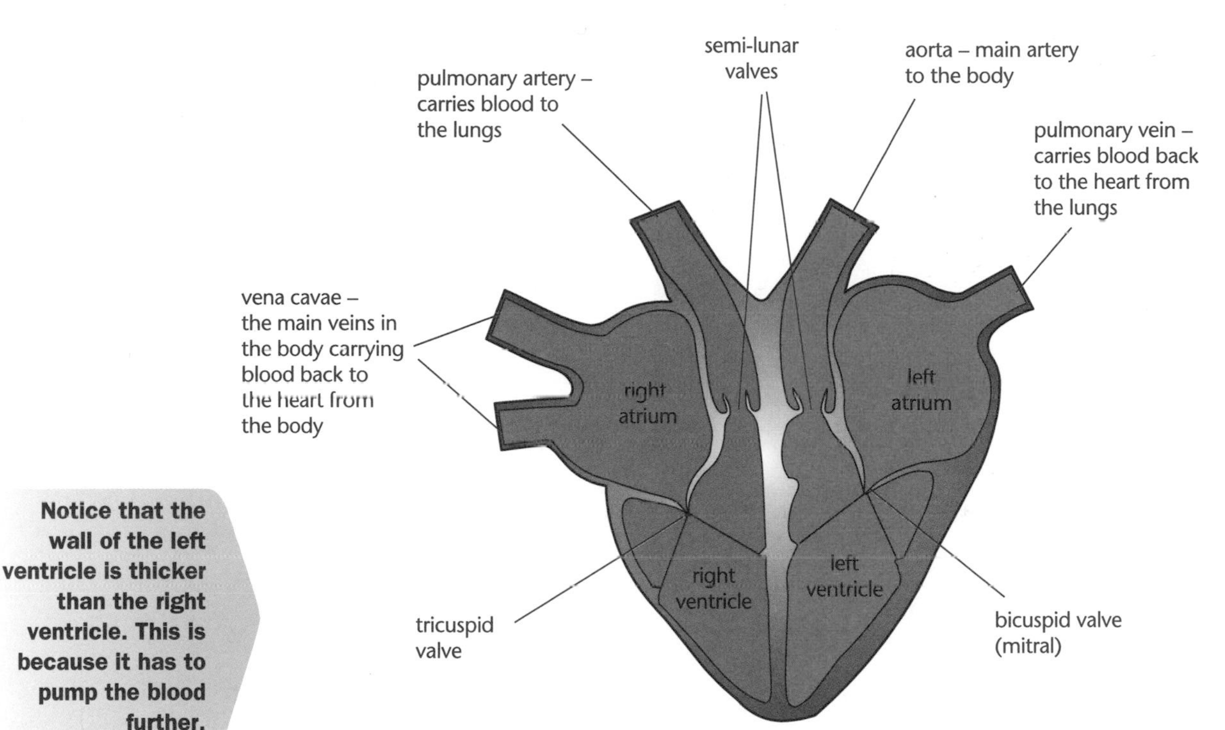

Notice that the wall of the left ventricle is thicker than the right ventricle. This is because it has to pump the blood further.

Fig. 7.2 Cross section of a heart.

Because the heart is such an important organ, any problems that occur are very serious. A type of fat called cholesterol can build up in the blood vessels that supply the heart muscle. This can cause areas of heart muscle to die. The heart valves may also be damaged by disease. Some people may need a new valve or even a new heart. Some artificial replacements are available but have drawbacks. Human organs for donation are in short supply and they may also lead to rejection problems.

Transport in plants

OCR B B3c

Plants have two different tissues that are used to transport substances. They are called **xylem** and **phloem**.

Xylem	Phloem
carries water and minerals from roots to the leaves	carries dissolved food substances both up and down the plant
the movement of water up the plant and out of the leaves is called **transpiration**	the movement of the dissolved food is called **translocation**
made of vessels which are hollow tubes consisting of thickened dead cells	made of columns of living cells

Water movement in plants

- Water enters the plant through the root hairs, by osmosis. It then passes from cell to cell, by osmosis until it reaches the centre of the root.
- The water enters xylem vessels in the root, and then travels up the stem.
- Water enters the leaves and evaporates. It then passes through the **stomata** by **diffusion**. This loss of water by transpiration helps to pull water up the xylem vessels.

How fast does transpiration happen?

The rate of transpiration depends on a number of factors:

- **temperature** – warm weather increases the kinetic energy of the water molecules so they move out of the leaf faster
- **humidity** – damp air reduces the concentration gradient so the water molecules leave the leaf more slowly
- **wind** – the wind blows away the water molecules so that a large diffusion gradient is maintained
- **light** – light causes the stomata to open and so more water is lost.

Why do plants lose water?

When plants are short of water, they do not want to waste it through transpiration. The trouble is they need to let carbon dioxide in, so water will always be able to get out. Water loss is kept as low as possible in several ways:

- Photosynthesis only occurs during the day, so the stomata close at night to reduce water loss.
- The stomata are placed on the underside of the leaf. This reduces water loss because they are away from direct sunlight and protected from the wind.
- The top surface of the leaf, facing the Sun, is often covered with a protective waxy layer.

Although transpiration is kept as low as possible, it does help plants by cooling them down and supplying leaves with minerals. It also provides water for support and photosynthesis.

7.3 Digestion

Digestion

AQA B2.11.6

Most **enzymes** work inside cells controlling reactions. Some enzymes pass out of cells and work in the digestive system. These enzymes digest our food, making the molecules small enough to be absorbed.

Details of the enzymes involved in digestion are shown on page 10.

7.4 Nerves, synapses and the brain

Nerves, synapses and the brain

OCR A B6.1–6

All living organisms need to respond to changes in the environment.

Although this happens in different ways, the pattern of events is always the same:

stimulus → detection → co-ordination → response

KEY POINT

The job of receptors is to detect the stimulus and effectors bring about the response.

It is the job of nerve cells to carry the messages around the body.

- **Sensory neurones** carry impulses from receptors to the CNS.
- **Motor neurones** carry impulses from the CNS to effectors.

Neurones pass impulses between each other across synapses. This process is described on page 15.

Ecstasy (MDMA) works by increasing the neurotransmitter **serotonin** in the brain. This increase in serotonin causes a change in mood.

Reflexes

Reflexes are responses that do not involve conscious thought. There are two main types.

Simple reflexes:

- produce a rapid response
- examples include the control of pupil size, many responses in newborn babies and much of the behaviour of simple animals.

Conditioned reflexes:

- this happens when an organism learns to link one stimulus with another stimulus, e.g. Pavlov's dogs

- the response therefore has no direct link with the stimulus
- they help organisms to avoid harm, e.g. birds avoid caterpillars that are a certain colour.

Memory and the brain

The brain contains billions of neurones and the **cerebral cortex** is the part of the brain involved with intelligence, memory, language and consciousness.

Scientists can work out the jobs of each part of the cortex by looking at patients with brain damage and using MRI scans.

Mammals **learn** from experiences by making new connections between neurones. This makes new nerve pathways which are more likely to be used in the future. This is the basis of **memories**.

People are more likely to remember things if:
- the stimulus is repeated a number of times
- if there is a pattern in the information
- if there is a strong stimulus, such as a smell, associated with it.

There are some skills that can only be learnt up to a particular age. Children can only learn some language skills up to a certain age.

HOW SCIENCE WORKS

OCR B B3c

Organ donation

Organ transplants involve the donation of organs from one person to another. Every year in the UK, about 2700 people receive a transplant. Transplants are sometimes the only treatment for certain patients. Attempts have been made to make mechanical replacements for organs, with little success.

NHS
Join the
NHS Organ Donor Register
Call 0845 60 60 400
www.uktransplant.org.uk

Kidney transplants are the most common, but transplants of the heart, liver and lungs are also regularly carried out. Other tissues such as corneas, heart valves, skin and bone can also be donated. As medical techniques have improved, more and more patients can be considered for transplants. This has led to a major problem, a serious shortage of donors.

For some people this means waiting for a long time and some may die before a suitable organ becomes available.

The line on the graph shows the number of people in the UK on the transplant list. The bars show the number of people who died and became donors and the number of transplants performed.

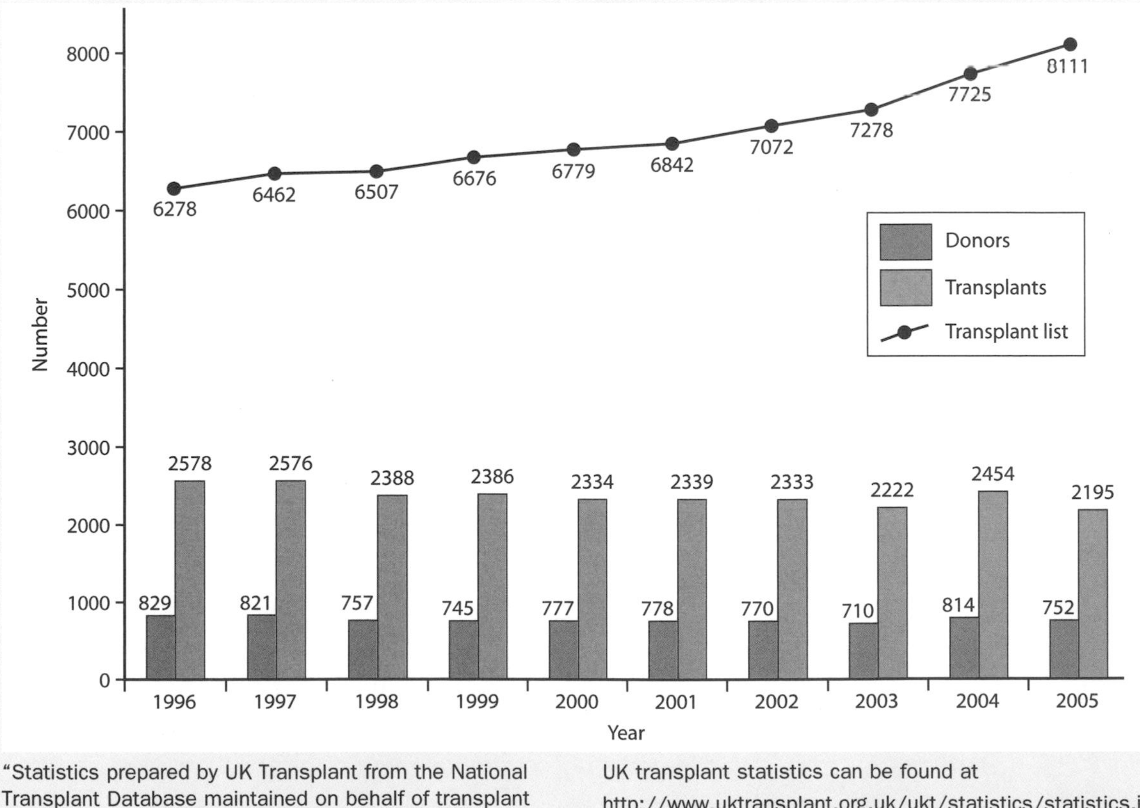

"Statistics prepared by UK Transplant from the National Transplant Database maintained on behalf of transplant services in the UK and Republic of Ireland."

UK transplant statistics can be found at http://www.uktransplant.org.uk/ukt/statistics/statistics.jsp
UK Transplant is part of NHS Blood and Transplant (NHSBT)

HOW SCIENCE WORKS

This article describes a recent idea to try and make more organs available for transplants.

A bid to boost organ donation has recently been rejected by MPs. The current system only allows organs to be removed if a person carries a donor card. Relatives of the person can stop the organs being taken even if the person is carrying a card. A recent survey showed 70% of British adults questioned are willing to donate their organs but many do not carry a card. The new system would mean that organs can be taken from a person unless they have opted not to allow it. This is called presumed consent.

Supporters of this new system say:

"This would greatly increase the numbers of organs available for transplants."

"When somebody dies it is very difficult to ask relatives about donation."

"In opt-out schemes already running, only 2% of people decide they do not want to donate their organs."

"The Presumed Consent scheme solves most of these problems. It forces people to make a decision."

People who are against the new idea say:

"Not everyone is happy with the idea of transplants. There are people who have religious and cultural objections."

"They might not realise that they need to opt out and could have their organs used."

"Mistakes could also be made and people could have their organs taken by mistake even if they have opted out. Getting it wrong could lead to distress for relatives and could lead to a backlash against doctors and organ donation."

HOW SCIENCE WORKS Questions

1. Look at the graph.
   (a) Describe what patterns it shows.
   (b) Explain how it highlights the need for more organs for transplants. [2]
2. Deciding about transplants involves a number of difficult ethical questions, what do you think?
   (a) Do you think the 'presumed consent' system should be introduced?
   (b) Is it right that many people receive organ transplants but do not carry donor cards themselves?
   (c) If there are a limited number of organs available, how do doctors decide which patients should receive transplants? [6]

Exam practice questions

1. The following structures are involved in transport in plants.
Match words **A**, **B**, **C**, and **D**, with the numbers **1–4** in the sentences.

A stomata
B xylem
C phloem
D root hairs

Plants take water up from the soil. Plants have many ____1____ to increase their surface area for water uptake.
The water is carried up the stem in the ____2____
Sugars, however, are transported in the ____3____
Water is lost to the air through ____4____ **[4]**

2. Put a tick (✓) next to any processes that are likely to occur in a person when their blood gets too hot.

Their muscles contract uncontrollably	
The blood vessels in the skin widen	
The sweat gland becomes less active	
The pituitary gland releases less ADH	

[2]

3. The boxes contain some blood vessels and some descriptions.
Draw straight lines to join each **blood vessel** to the correct **description**.

Blood vessel	**Description**
aorta	carries oxygenated blood under low pressure
pulmonary artery	carries blood into the right atrium
pulmonary vein	carries deoxygenated blood away from the heart
vena cava	carries oxygenated blood under high pressure

[3]

Exam practice questions

4. The graph shows the levels of glucose and insulin in Alysha's blood.
The readings were taken for 60 minutes after Alysha had eaten a meal of rice.

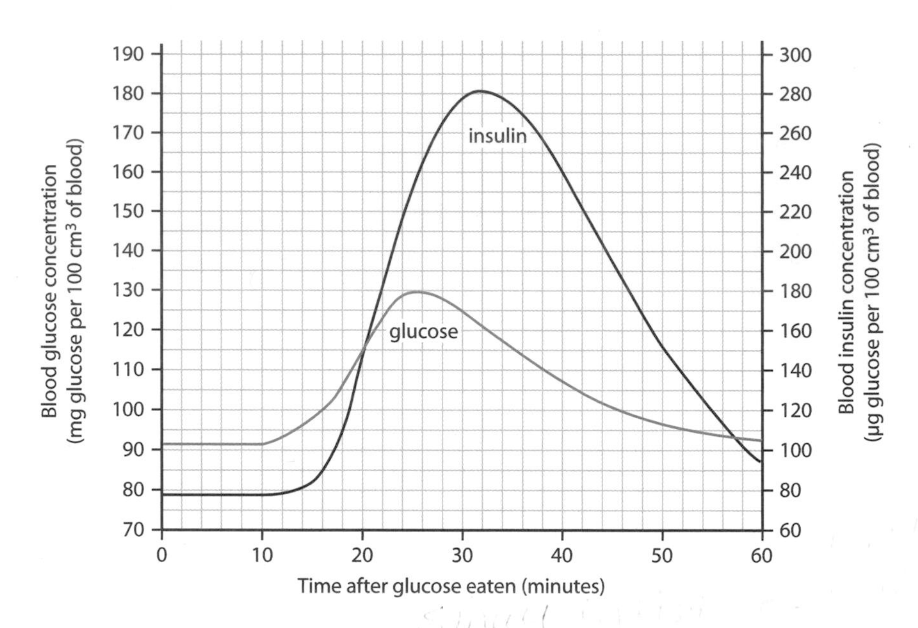

(a) Rice contains a high quantity of starch.
Describe how and where this starch is digested in the body. **[3]**

(b) **(i)** How long does it take for the blood glucose levels to rise after the meal? **[1]**
(ii) Give two reasons why it takes some time for the glucose levels to rise. **[2]**

(c) **(i)** What is the maximum level of insulin in the blood during the experiment? **[1]**
(ii) Explain why insulin levels rise after the meal. **[3]**

(d) How would you expect the two lines on the graph to be different if Alysha has diabetes? **[2]**

5. Read the How Science Works pages about organ donation and answer the following questions.

(a) The article states that it is difficult to produce mechanical replacements for many organs.
Suggest one reason why. **[1]**

(b) The table contains some arguments about the idea of presumed consent.
Put a (✓) next to any statements that could be used to argue for presumed consent.

Statement	
Some people's organs could be removed in error	
It would increase the number of organs available for transplants	
It would avoid having to ask relatives about donation	
Some people have religious objections to transplants	

[2]

(c) According to the article, 70% of people in the UK say that they are willing to donate organs but do not carry donor cards.
Suggest two reasons why this might be. **[2]**

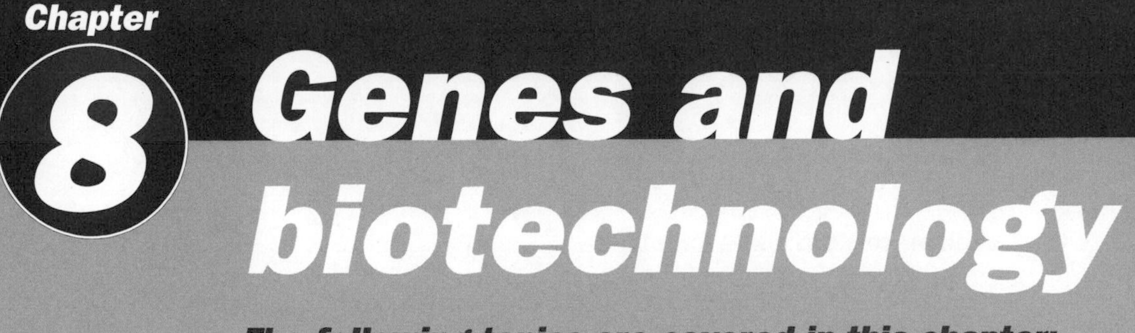

Chapter 8 Genes and biotechnology

The following topics are covered in this chapter:

- ***Using our knowledge of genes***
- ***Biotechnology***

8.1 Using our knowledge of genes

Genetic fingerprints

B3a
B2.11.8

Scientists have discovered that our DNA contains regions that do not code for proteins. This is often called junk DNA. In this DNA are regions with repeating sequences that can be used to identify each individual.

A genetic fingerprint is a pattern of DNA that can be used to identify an individual.

Step 1 – the regions of the DNA are isolated and cut up using enzymes.

Step 2 – the DNA fragments are put on a gel and separated using an electric current.

Step 3 – the fragments are treated with a radioactive probe so that the bands of DNA can be photographed.

Selective breeding

B2.11.8

Selective breeding involves choosing animals or plants with desired characteristics and allowing them to breed.

Selective breeding is not new; farmers have been altering the genes of their plants and animals for thousands of years. For example, to produce larger hens' eggs, farmers bred the hens that produced the largest eggs, with cocks hatched from large egg laying mothers. They repeated this for several generations. Other examples include: breeds of dogs, higher yielding crops with better flavour and resistance to disease.

There are problems with selective breeding:

- the animals all become too closely related or inbred
- the process can take a long time.

A quicker way of changing the genes of animals or plants is **genetic engineering**.

Genetic engineering

OCR B B3g
EDEXCEL 360 B2.1–2

KEY POINT **Genetic engineering involves taking DNA from one organism and putting it into the chromosomes of another organism.**

When scientists investigated the genetic code, they found that all living organisms use the same code. This makes it possible to move genes from one organism to another and it can even be to an organism of another species; from a person to a bacterium, for example.

A number of steps in the process involve using enzymes:

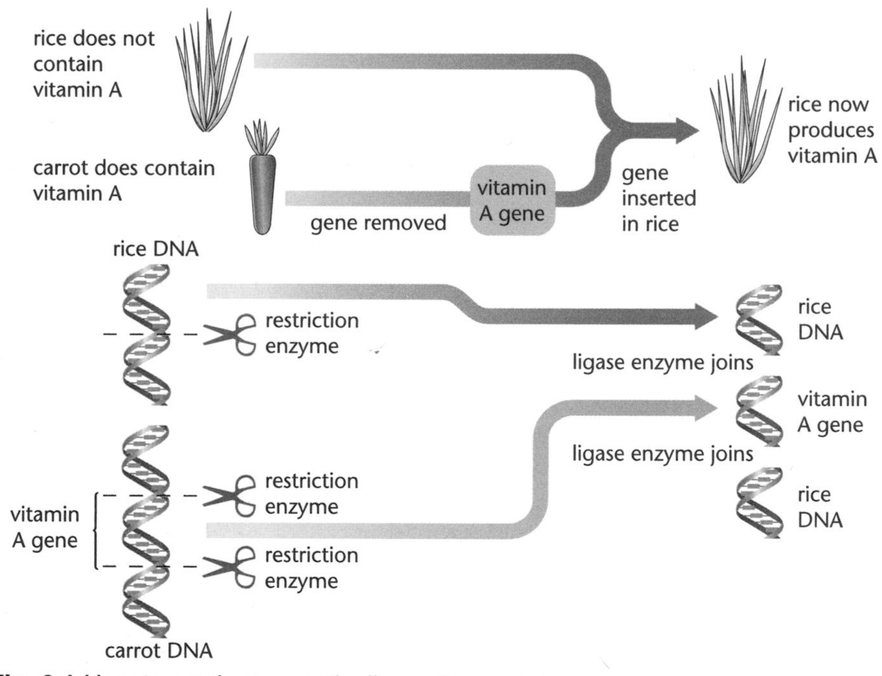

Fig. 8.1 How to produce genetically engineered rice.

8.2 Biotechnology

Useful microorganisms

Microorganisms have been used for hundreds of years for making foods such as bread and cheese. They are now being used more and more to produce new types of food and other useful products.

KEY POINT

The use of living organisms to make useful products is called biotechnology.

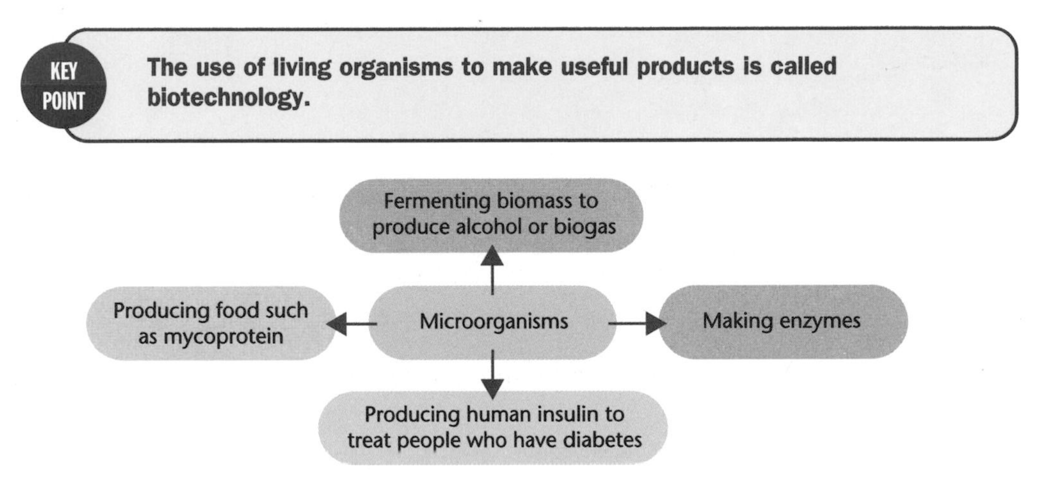

Microorganisms can be grown in large vessels called **fermenters**. The conditions are carefully controlled so that the microorganisms grow very fast. Like all organisms, they need food to obtain energy but they can often be fed on waste from other processes.

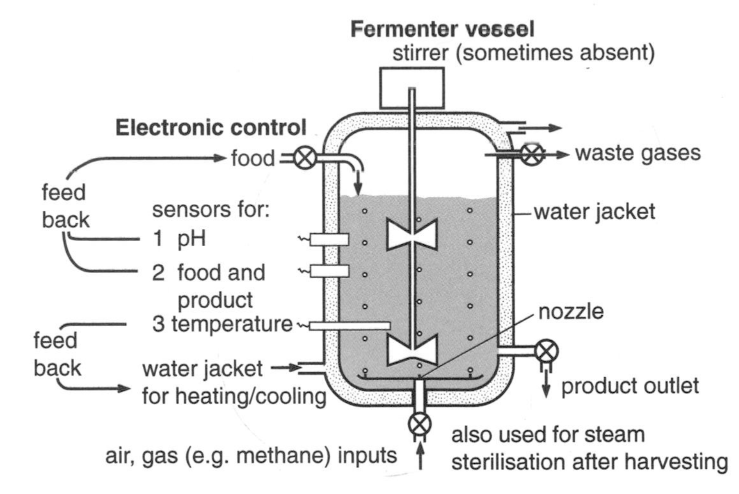

Fig. 8.2 A fermenter.

The microorganisms can be fed on **biomass** such as manure or plant material. The ethanol or biogas that is produced can be used as a fuel.

The **enzymes** that some microorganisms make can be used for a number of different processes:

- protein and fat digesting enzymes (proteases and lipases can be used in washing powders)
- carbohydrases can be used to convert starch into sugar syrup
- isomerase can be used to convert the sugar glucose into fructose which is much sweeter.

HOW SCIENCE WORKS

OCRB B3g
EDEXCEL 360 B2.2

Vitamin A

Vitamin A is a fat-soluble vitamin. The best sources of vitamin A are eggs, milk, butter and fish. Plants do not contain vitamin A, but they contain beta-carotene. Beta-carotene can be converted to vitamin A in the body. The best sources of beta-carotene are dark-green, orange, and yellow vegetables. Cereals are poor sources of beta-carotene.

Vitamin A is used for two functions in the body. In the eye, it is needed to make light sensitive cells for seeing in dimlight. Vitamin A is also important during the development of the embryo.

Vitamin A deficiency is common in areas like Southeast Asia, where polished rice, which lacks the vitamin, is a major part of the diet. The earliest symptom of vitamin A deficiency is night blindness. Continued vitamin A deficiency leads to inflammation and infection in the eye, resulting in total and irreversible blindness. Vitamin A deficiency can be prevented or treated by taking vitamin supplements or eating food rich in vitamin A. In Africa, Indonesia, and the Philippines, programs supply children with injections of the vitamin. However, figures show that 500 000 children go blind each year due to vitamin A deficiency.

It has also been shown that children can be born with a number of different birth defects if their mother lacked vitamin A when she was pregnant.

So should everybody take vitamin A tablets?

We have to be careful because studies have shown that too much vitamin A may be dangerous. There have been lots of different studies, but some studies are better designed than others.

Look at this study:

- In a recent study, scientists questioned over 22 000 women.
- They asked them about their eating habits and which vitamin tablets they took before and during pregnancy. This was used to see how much vitamin A they ingested.
- They then counted the numbers of any birth defects in the babies.
- They concluded that vitamin A was dangerous because the total number of different defects increased as vitamin A intake increased.

Golden rice

Due to the number of people with vitamin A deficiency, scientists have been developing genetically modified rice rich in beta carotene. The idea is that this will help people who can't afford a diet containing enough natural sources of vitamin A. The golden rice project is one project and is already undergoing trials. Different groups have different views on this project.

HOW SCIENCE WORKS

"I think that the introduction of golden rice would be a real mistake. A person would have to eat about twelve times the amount of rice that they eat now to have any effect. I also think that it is wrong to introduce this rice because it might make people dependant on one food only. They might start to suffer from other deficiency diseases. It would be better to spend money introducing many vitamin-rich food plants that are cheap and already available.

Anyway I think there are dangers with any genetically introduced crops. We do not really know how safe the Golden Rice is. It could also breed with wild relatives to contaminate wild rice forever. If there were any problems, the clock could not be turned back."

"I think that the use of golden rice is a real step forward. We have tried other methods but there are still 500 000 children a year going blind. It is time to try something new. If this method works then the gene could be put into other crops like sweet potatoes and cassava. If it works with vitamin A then genes for other vitamins or amino acids could be introduced. I don't think that they will have to eat large amounts to see an effect. The figure of twelve times the dose may be an ideal amount but improvements will be seen by eating smaller amounts."

HOW SCIENCE WORKS Questions

When any new scientific study is performed it should be a **fair test** and obtain data that is **accurate** and **reliable**.

Look at the study on vitamin A and pregnant women.

(a) Do you think it was a fair test? **[1]**

(b) Do you think the method for measuring vitamin A intake was accurate? **[1]**

Exam practice questions

1. The following processes involve manipulating genes.
Match words **A**, **B**, **C**, and **D**, with the numbers **1–4** in the sentences.

A genetic engineering
B gene therapy
C selective breeding
D genetic fingerprinting

Farmers have been manipulating the genes of their animals by choosing which animals mate. This is called ___1___
It is now possible to make bacteria produce human insulin by the process of ___2___
If a person's genes are altered to cure a disease this is called ___3___
A scientist may use ___4___ to identify a person from a sample of their cells. **[4]**

2. The diagram shows some stages in the production of human insulin by bacteria.

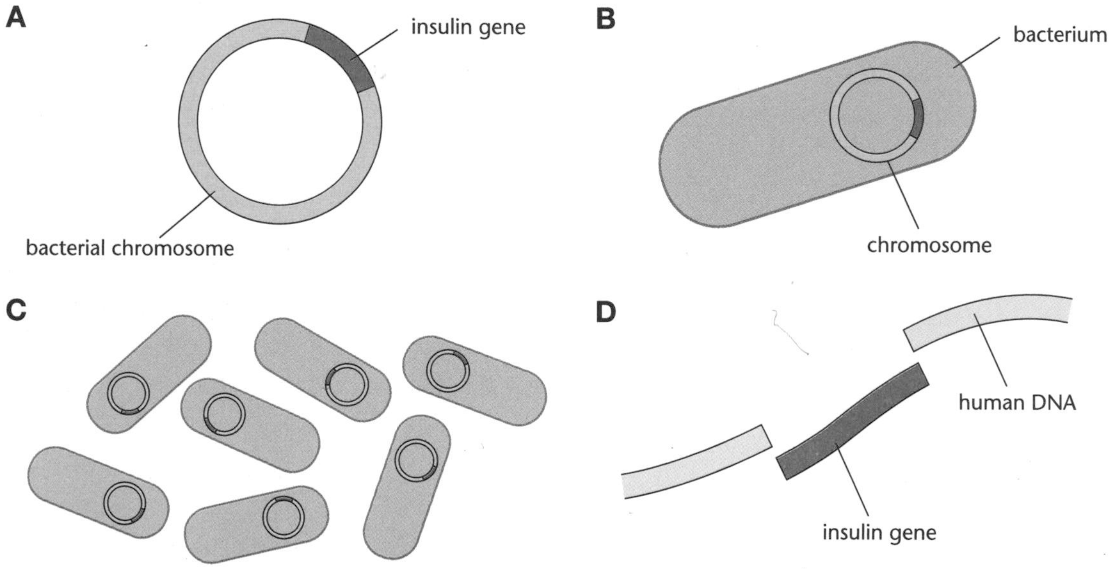

A
insulin gene
bacterial chromosome

B
bacterium
chromosome

C

D
human DNA
insulin gene

(a) Write down the order that the steps would take place.
The first one has been done for you.

D ______ ______ ______ **[2]**

(b) The table contains some statements about this method of making insulin.
Put an (✗) next to any incorrect statements.

The process uses hormones to cut DNA	
The bacteria can be grown in large fermenters	
The insulin produced is a hybrid of human and bacterial insulin	
The bacteria can be grown on cheap waste products	

[2]

3. **(a)** The boxes contain some parts of a fermenter and their jobs.

Exam practice questions

Draw straight lines to join each part to its correct job.

Part	Job
steam inlet	to keep a constant temperature in the fermenter
water jacket	to mix the microorganism with the food
stirrer	to allow the microrganism to respire
air inlet	to sterilise the fermenter between batches

[3]

(b) Fermenters that produce biogas are often used in remote areas of countries.

(i) Write down **one** raw material that can be used to produce biogas. [1]

(ii) Write down **two** advantages of using biogas as a fuel, rather than fossil fuels. [2]

4. Read the How Science Works article and use it to answer the following questions.

(a) Look at these possible changes to the vitamin A and pregnant women study.

A Question more than 22 000 women.
B Measure vitamin A intake by taking regular blood samples.
C Study women in different countries.

(i) Which of these changes would make the data from the vitamin A and pregnant women study more **accurate**? [1]

(ii) Which of these changes would make the data from the vitamin A and pregnant women study more **reliable**? [1]

(b) The study judged the danger of vitamin A by simply counting the number of any defects.
Is this a valid way of judging the danger of high vitamin A levels? [2]

(c) The two people have different views on how much golden rice a person needs to eat to have an effect.
Why do you think their views differ? [2]

More about microbes

The following topics are covered in this chapter:

- *Bacterial structure and growth*
- *Microorganisms and food*
- *Microbes and genetic engineering*
- *Microorganisms and disease*
- *Biofuels*

9.1 Bacterial structure and growth

Bacterial structure

Bacteria can feed in many different ways.

- Some are **parasites** and feed on living organisms causing diseases.
- Others are **decay** organisms feeding on dead organic material.
- Some can **photosynthesise** and make their own food.

These differences allow them to live in many places and feed on many different things.

They also have many different shapes.

Bacteria are often named according to their shape.

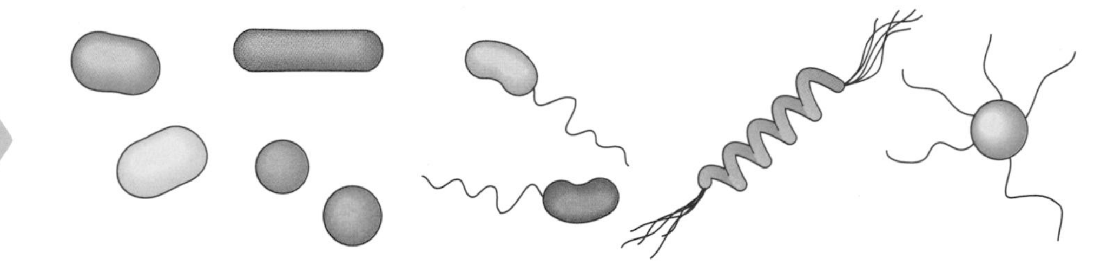

Fig. 9.1 Different types of bacteria.

Despite these differences all bacterial cells have a similar structure.

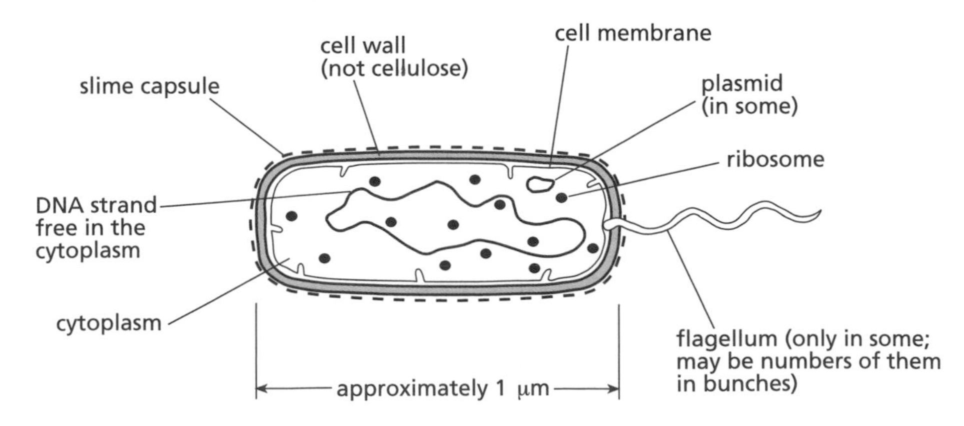

Fig. 9.2 A generalised bacterium.

Bacterial cells are smaller than animal and plant cells. They lack a true nucleus, mitochondria, chloroplasts and vacuoles.

Growing bacteria

AQA B3.13.5

Up until the late 1700s, most people thought that living organisms could just appear in non-living material. Several scientists were involved in disproving this idea:

- Lazzaro **Spallanzani** in 1768 showed that bacteria only appear in meat broth if the air could get in.
- Theodore **Schwann** in 1839 put forward the cell theory.
- Louis **Pasteur** in 1860 improved on Spallanzani's experiment and finally convinced people that life cannot come from non-living material.

The idea that life can only be passed on from living organisms is called biogenesis.

It is possible to grow microorganisms such as bacteria in laboratories. They are grown on a special jelly called an **agar** in a Petri dish. The agar is a culture medium containing an energy source, minerals and sometimes vitamins and protein.

Precautions have to be taken:

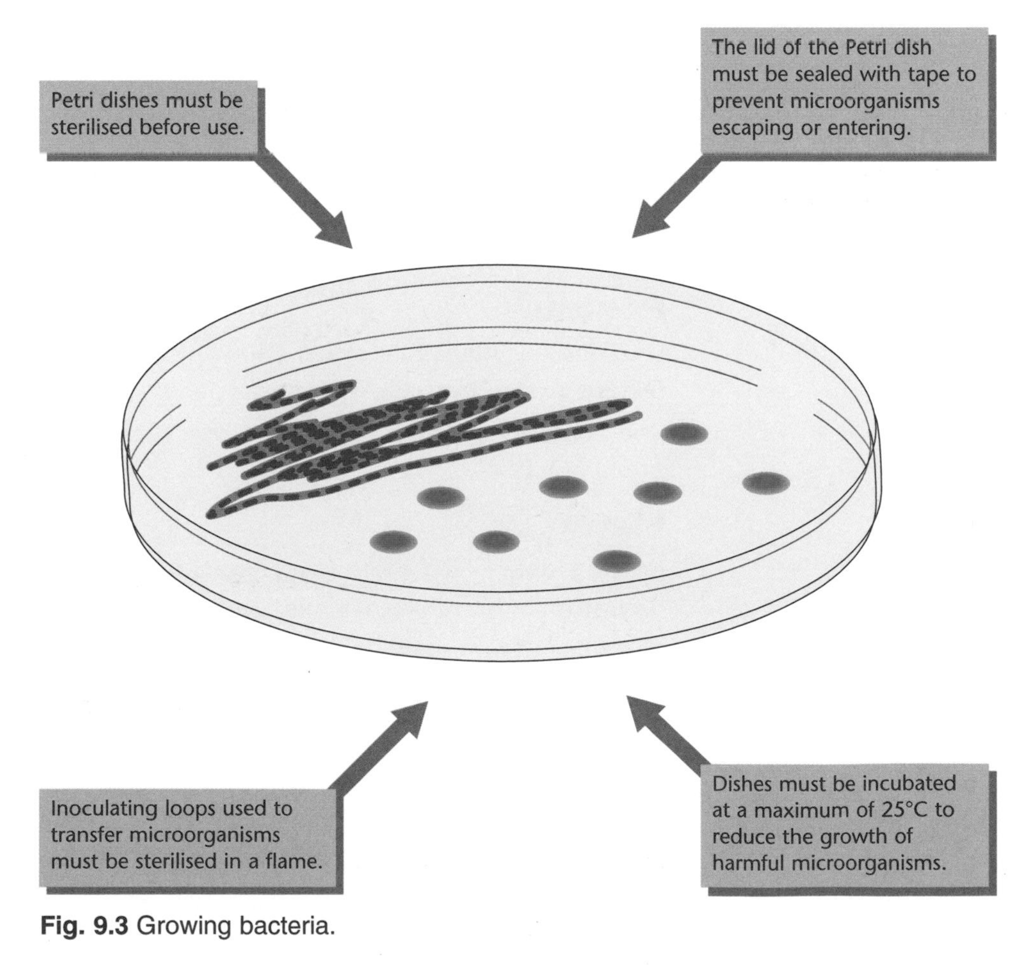

Fig. 9.3 Growing bacteria.

9.2 Microorganisms and disease

Pathogenic microorganisms

OCR B B6B

Different types of organisms can cause various diseases.

They can enter the body in different ways.

Type of organism	Example	Entry route	Disease caused
bacteria	Salmonella, E. coli Vibrio	in food in water	food poisoning cholera
protozoa	Entamoeba	in water	dysentery
virus	Adenovirus	breathed in	influenza

Organisms that cause disease are called pathogens.

For a pathogen to cause disease, various steps usually happen:

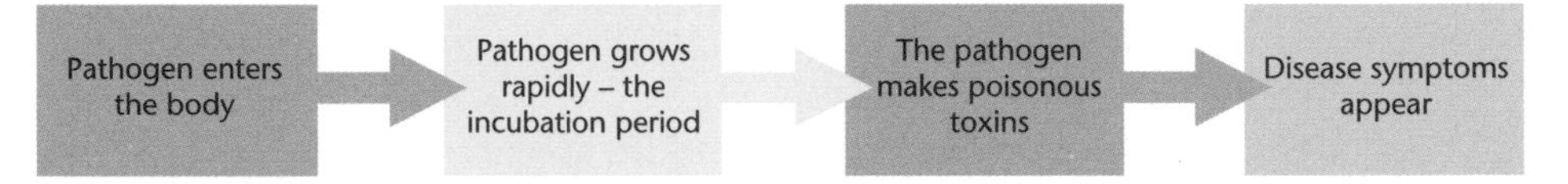

These diseases can occur at any time, but after a natural disaster large numbers of people may become ill. This is because sewage and water systems may become contaminated and refrigerators may not work due to lack of electricity.

Modern medicine can treat many diseases. This is the result of discoveries made by many scientists. Pasteur, Lister and Fleming are three of these scientists.

Pasteur

Louis Pasteur studied a number of diseases such as rabies and anthrax. He was the first person to realise that diseases that can be passed on are caused by living organisms. This is called the **germ theory**.

Lister

Joseph Lister worked as a doctor, operating on patients. He found that treating his instruments and washing his hands with a chemical called carbolic acid helped to stop his patients' wounds becoming infected. This was the first **antiseptic** to be used.

Fleming

Sir Alexander Fleming was growing bacteria on Petri dishes. He noticed that one of his dishes had a fungus growing on it. Around the fungus, called *Penicillium*, the bacteria had been killed. The fungus was producing a chemical called penicillin which was the first **antibiotic** to be discovered.

Two other scientists were responsible for isolating penicillin from the fungus.

9.3 Microorganisms and food

Traditional food and drink

OCR B B6c
AQA B3.13.4
EDEXCEL B3.1

Microorganisms have been used for many years to produce different types of food and drink. All these processes involve a particular microbe and use a particular energy source. The microbe respires and makes the product.

Energy source + **Microbe** —Respiration→ **Products**

Yoghurt

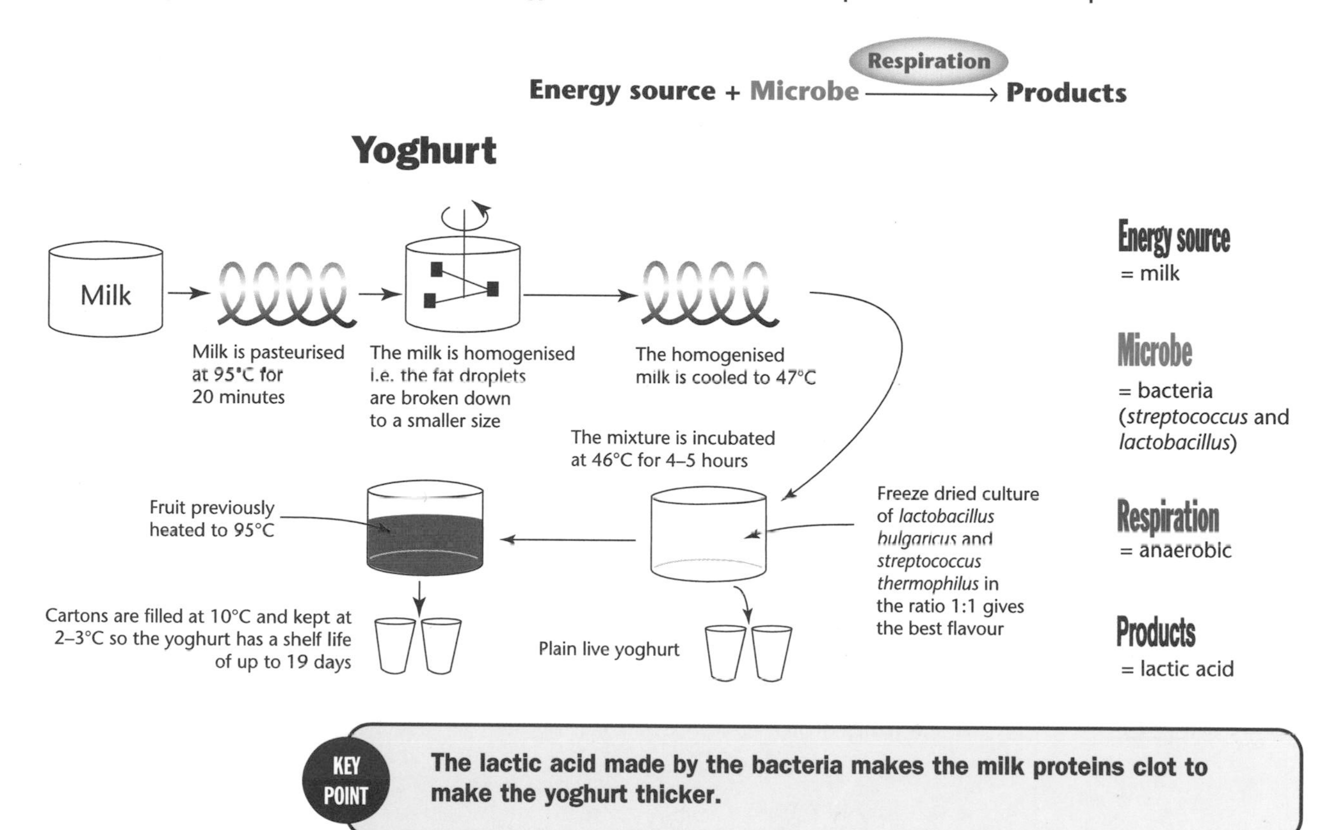

KEY POINT
The lactic acid made by the bacteria makes the milk proteins clot to make the yoghurt thicker.

Alcohol

Energy source = grapes for wine or barley for beer
Microbe = yeast
Respiration = anaerobic
Products = alcohol and carbon dioxide

Making alcohol relies on yeast fermenting sugar. This sugar can come from different sources. The equation for fermentation is:

glucose	→	ethanol (alcohol)	+	carbon dioxide
$C_6H_{12}O_6$		$2C_2H_5OH$		$2CO_2$

In beer making, barley is allowed to germinate so that the starch is turned into sugar. Hops are added to give the beer flavour. After the yeast has fermented, the beer is drawn off the yeast. It may then be heated (pasteurised) to kill any microbes before it is bottled.

The concentration of alcohol made by fermentation is limited. This is because when the alcohol reaches a certain concentration it will kill the yeast.

Wine is not usually any stronger than about 15% alcohol.

Spirits are therefore made by **distillation**. This concentrates the alcohol.

The type of spirit that is produced depends on the sugar source:

- rum from sugar cane
- whisky from malted barley
- vodka from potatoes.

Soya sauce

Energy source = soya beans
Microbe = fungus (*Aspergillus*) then yeast and bacteria.
Respiration = anaerobic
Products = soya sauce

Modern uses of microbes for food

Mycoprotein

Energy source = any starch or sugar source such as potato waste
Microbe = a fungus (*Fusarium*)
Respiration = aerobic
Products = mycoprotein

Mycoprotein is often used as a meat substitute as it has a high protein content. It does have a number of advantages over meat:

- the fungus grows very quickly
- it has a high fibre content
- it is low in fats
- it can be grown on waste substances.

9.4 Biofuels

Going green

KEY POINT

Biofuels are fuels that have been made from living material that has been produced in a sustainable way.

Fossil fuels such as coal and oil have been produced from living material over millions of years. They are not sustainable as we are burning them faster than they are being produced. This means that carbon dioxide is being added to the air.

Microbes can be used to produce biofuels such as **biogas** and **ethanol**. These fuels are sustainable because they release carbon dioxide to the air at the same rate that it is being absorbed by photosynthesis.

Biogas

This is produced in fermenters called **digesters**.

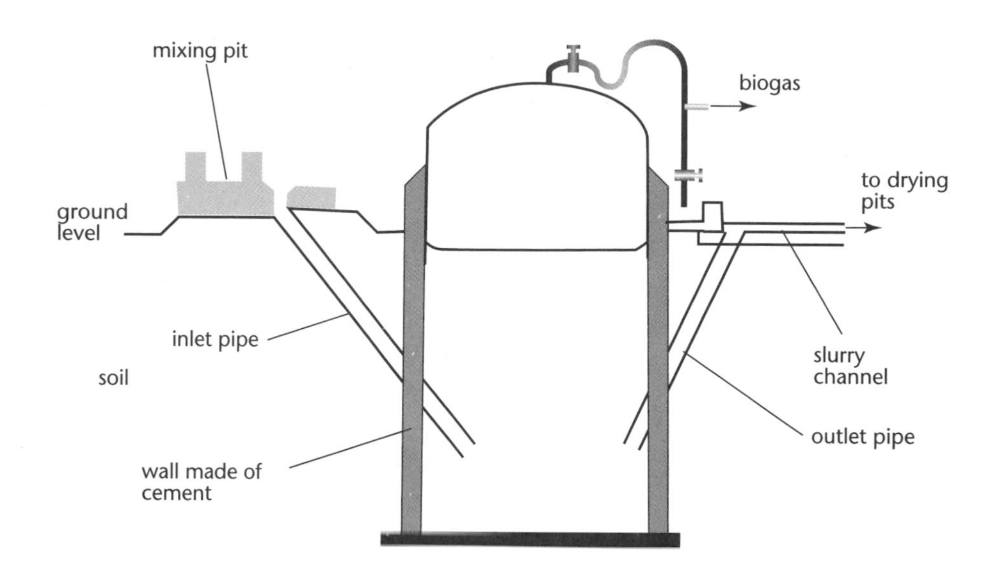

Fig. 9.4 A biogas digester.

Waste materials such as sewage or plant products are put in the tank. A mixture of different bacteria use these substances for anaerobic respiration. They produce biogas which is made up of:

Too little methane and the gas is explosive!

- largely methane
- some carbon dioxide
- small amounts of hydrogen, nitrogen and hydrogen sulfide.

The biogas can then be burnt to make electricity, produce hot water or power motor vehicles.

Ethanol

Some countries grow large amounts of sugar cane or maize. These crops can be used to produce sugar that can be fermented by yeast. This produces ethanol.

The ethanol can be concentrated by distillation and added to petrol.

Petrol (about 85%) + **Ethanol (about 15%)** ⟶ **Gasohol**

Cars can be converted to run on the gasohol and so less petrol is used.

9.5 Microbes and genetic engineering

How is genetic engineering done?

OCR A B7.4
OCR B B6h

It is now possible to make microbes produce different proteins by changing their DNA. They are then called **genetically modified** or **GM**.

If the gene has come from a different species then the GM organism is called transgenic.

The main steps in making a GM organism involve:

To make GM bacteria the vector is usually a ring of DNA called a plasmid.

- finding and removing the necessary gene
- putting the gene into a vector that will carry it into the new cell
- testing the cells to see if they have taken up the gene
- allowing the new cells to reproduce and produce the new protein.

The whole process is controlled by enzymes and is shown in Fig. 8.1 on page 104.

GM organisms can have lots of possible uses:

GM microbes	GM plants	GM animals	GM humans
making chymosin	resistance to weedkillers	faster growth	gene therapy
making insulin	increasing yield	producing useful proteins in milk	
	produce insecticide		
	resistance to disease		

Some of the arguments for and against genetic engineering are on pages 42–43.

Insulin and chymosin production

OCR A B7.4
OCR B B6H

Insulin

Insulin is needed to treat people who have diabetes. This is explained on page 18. Until 1982, the insulin used was extracted from cows or pigs. This was difficult to extract and worked slower than human insulin.

Now, the human gene for insulin is placed into bacteria and they produce insulin that is identical to human insulin. This works faster and can be produced in large quantities.

Chymosin (rennin)

Chymosin is an enzyme that has been used in cheese making for many years. It is produced in the stomach of young mammals and makes milk proteins clot.

To make cheese, it has been necessary to kill animals and extract the chymosin from their stomach.

Now most chymosin is made by GM microbes, usually fungi. They have had the gene for chymosin inserted. This means that large amounts of chymosin can be made quickly and the cheese can be eaten by vegetarians.

Herbicide and insect resistance

EDEXCEL B3.1

It is possible to use genetically modified bacteria to produce GM plants. The bacteria are used to inject useful genes into plants.

The type of bacteria used (*Agrobacterium tumefaciens*) injects its genetic material into plant cells producing a growth of plant cells called a gall. The cells of this gall can be grown into a complete GM plant.

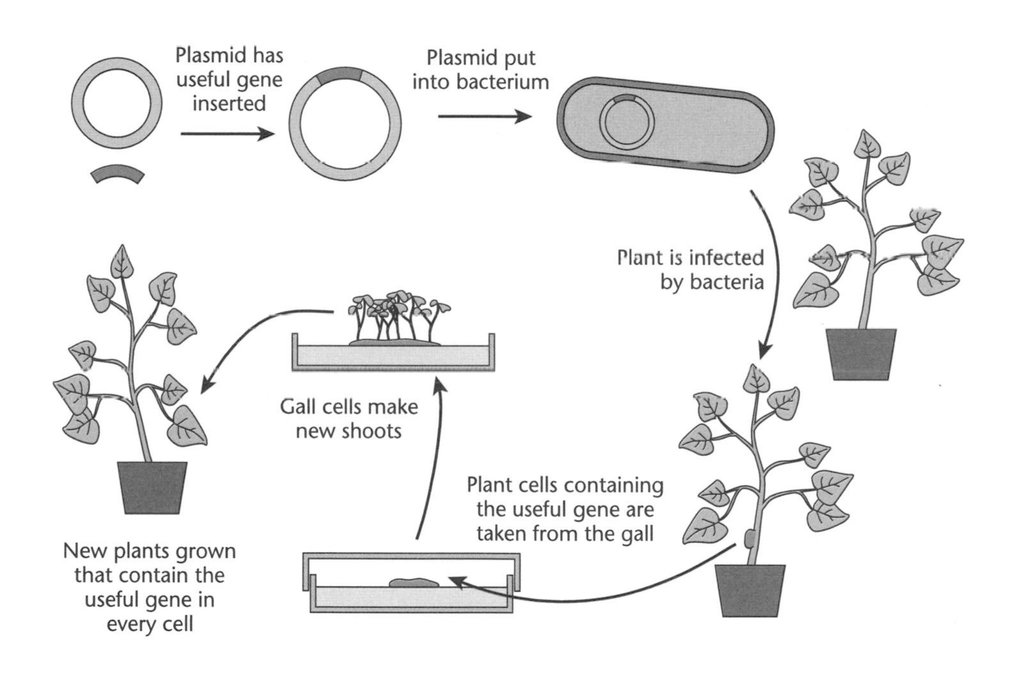

Fig. 9.5 Producing a GM plant.

Several different genes can be inserted:

- A gene may make the plant resistant to weedkillers (herbicides). This means the farmer can spray his fields to kill weeds without killing the crop.
- A gene from another type of bacteria (*Bacillus thuringiensis*) can be inserted. This makes the crop plant make its own insecticide.

HOW SCIENCE WORKS

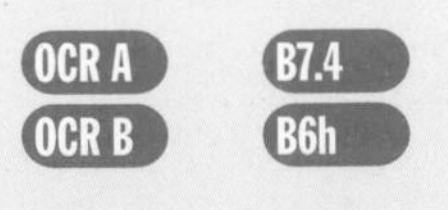

Is human insulin better than animal insulin?

The use of insulin injections to treat people who are diabetic was first developed in the 1920s by two scientists called Banting and Best. They performed experiments on dogs. By removing the pancreas of dogs they made the dogs diabetic. They then took a pancreas from another dog and minced it up. By injecting an extract into the diabetic dogs they made them better. They then set about extracting enough insulin to use on a person.

In 1922 they experimented on a young boy called Leonard Thompson. He was 14 years old and close to death due to diabetes. Thanks to the injections, he recovered and lived for another 13 years. Since then the lives of millions of people have been saved by injections of animal insulin. In 1982 genetically modified insulin, 'human' insulin was developed. It was the first genetically produced drug to be licensed and used on people.

Everybody assumed that this insulin would be better for diabetics than animal insulin. But is this really true? It certainly has slightly different effects.

Insulin type	Time taken to start to work in hours	Peak effect in hours	Length of time it works for in hours
ANIMAL	0.5 – 2	3 – 4	4 – 6
'HUMAN'	0.5 – 1	2 – 3	3 – 6

Most diabetics now receive human insulin but there is some argument about whether animal insulin should still be used.

"GM 'human' insulin must be better than animal insulin. It is an exact copy of the insulin made by the body. This means that the body will produce less antibodies. It is also cheaper, therefore it is more available to people in poor countries. Because it works quicker I think that the blood sugar level will be controlled better and this will cause less damage to the body."

"There is no evidence that shows that human insulin is better than animal insulin. No large scale, long-term trials have ever been carried out. Human insulin can cause side effects for some people and these effects often disappear if they change back to animal insulin. I think diabetics should be given a choice."

The next step?

A Canadian company have managed to put the human insulin gene into a plant.

They are growing safflower which is a plant that is sometimes grown for the oil in its seeds. The genetically modified plants now produce insulin in their seeds. The company think that one large farm could make all the insulin that the world needs. It would be much cheaper than growing bacteria in fermentation tanks.

Some people are worried because these plants are being grown out in the open.

HOW SCIENCE WORKS *Questions*

1. Very few people objected to Banting and Best's experiments on dogs.
   (a) Why do you think this was? **[2]**
   (b) Why do you think they were so keen to give Leonard Thompson the extract even though it had not been thoroughly tested? **[1]**
2. Do you think animal insulin should still be available for use by diabetics? **[2]**
3. Why do you think that some people are worried about growing crops that contain insulin? **[2]**

Exam practice questions

1. The following terms can refer to bacteria.
Match words **A, B, C**, and **D**, with the numbers **1–4** in the sentences.

A photosynthesis
B decomposers
C host
D pathogens

Many bacteria feed on dead material. They are called ____**1**____.
Others may cause disease and are called ____**2**____.
The living organism that they feed on is called the ____**3**____.
Other bacteria may make their own food by ____**4**____. **[4]**

2. Put a tick (✓) or a cross (✗) in each empty box to complete this table about the structure of bacteria and animal cells.

Feature	**Bacteria**	**Animal cell**
They have a cell membrane		
They all have identical shapes		
They have mitochondria		
Their DNA is kept in a nucleus		

[4]

3. The boxes contain some famous scientists and descriptions of biological developments.
Draw straight lines to join each **Scientist** to the **Development** that they were most responsible for.

Scientist	**Development**
Alexander Fleming	The development of the germ theory
Joseph Lister	The discovery that certain fungi produce antibiotics
Louis Pasteur	The development of the cell theory
Theodore Schwann	The use of antiseptics

[3]

Exam practice questions

4. Look at the diagram on page 117 which shows a digester used to make biogas.

(a) Suggest **two** reasons why the digester is buried in the soil with a thick concrete wall. [2]

(b) An input of air to the digester is not needed. Explain why. [1]

(c) Explain why biogas is considered to be a sustainable fuel. [2]

(d) Why is biogas production so useful in remote areas in countries such as India? [2]

5. Danny was making some wine using some fruit juice, yeast and sugar.

(a) He uses a glass jar.

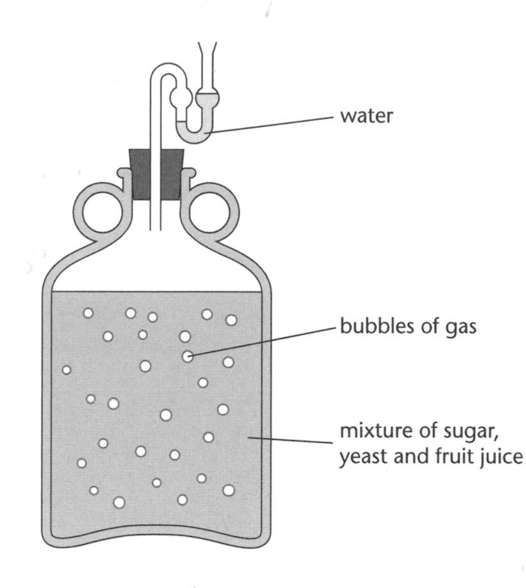

(i) What is the main gas found in the bubbles in the jar? [1]

(ii) Why does he use an S-shaped tube in the neck of the jar? [2]

(b) Danny wants to see if he could make wine with more alcohol by adding twice the amount of sugar than his recipe suggested. He sets up two jars with the same amount of yeast and fruit juice but one with extra sugar. He then measures the concentration of alcohol in the wine every day.
His results are shown in the graph.

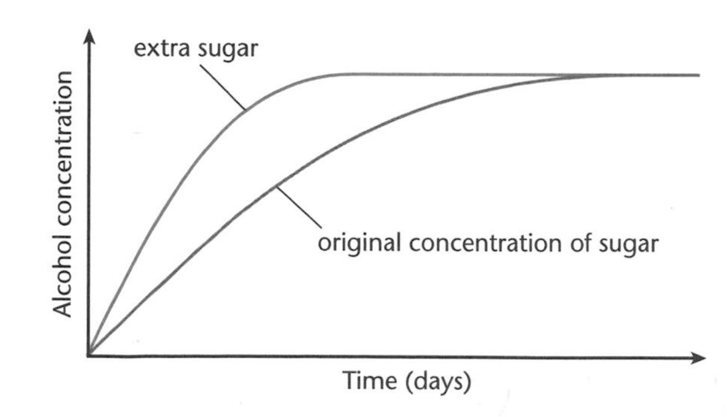

(i) Danny used the same amount of yeast and fruit juice in each of the two jars. Write down **one** other thing he should do to make his investigation a fair test. [1]

(ii) Describe what his graph shows about the differences between the fermentation in the two jars. [2]

(iii) Suggest explanations for his results. [2]

(iv) What difference do you think there would be in the taste of his two wines? [1]

Use, damage and repair

The following topics are covered in this chapter:

- ***The skeleton and exercise***
- ***Respiratory systems***
- ***Reproduction and growth***
- ***The circulation***
- ***Removing waste***

10.1 The skeleton and exercise

The organisation of the skeleton

OCR A B7.7
OCR B B5a

Different animals have different types of skeletons:

- animals, like insects, have an **external** skeleton
- all vertebrates have an **internal** skeleton. This makes it easier to grow and to attach muscles to.

The skeleton carries out important functions.

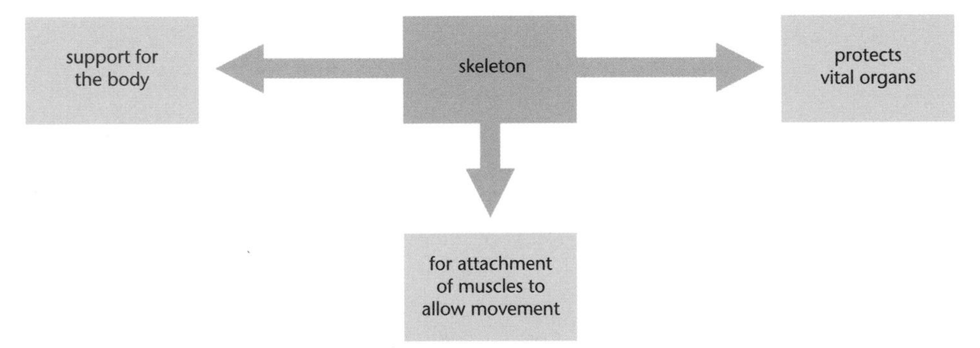

The skeleton of vertebrates is made up of **bone** and **cartilage,** both of which are living tissues. In animals like sharks, the skeleton is mainly cartilage but in us it is mainly bone.

The long bones of our arms and legs are hollow. This makes them much lighter but still strong.

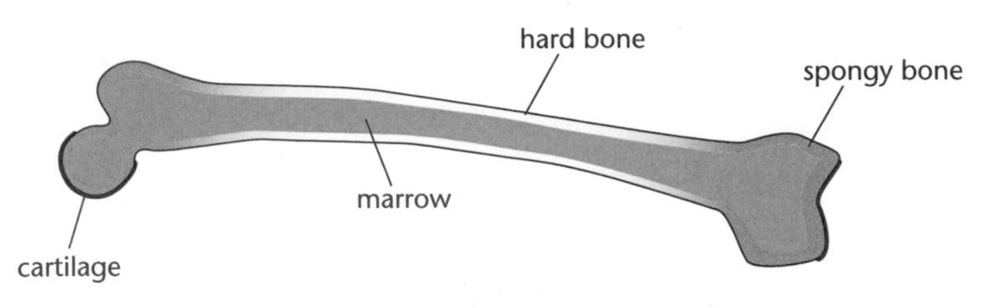

Fig. 10.1 The structure of a long bone.

The bone marrow makes blood cells.

Joints

Where two bones meet they form a **joint**. Some of these joints are fused but others allow movement.

Synovial joints are specially adapted to allow smooth movement.

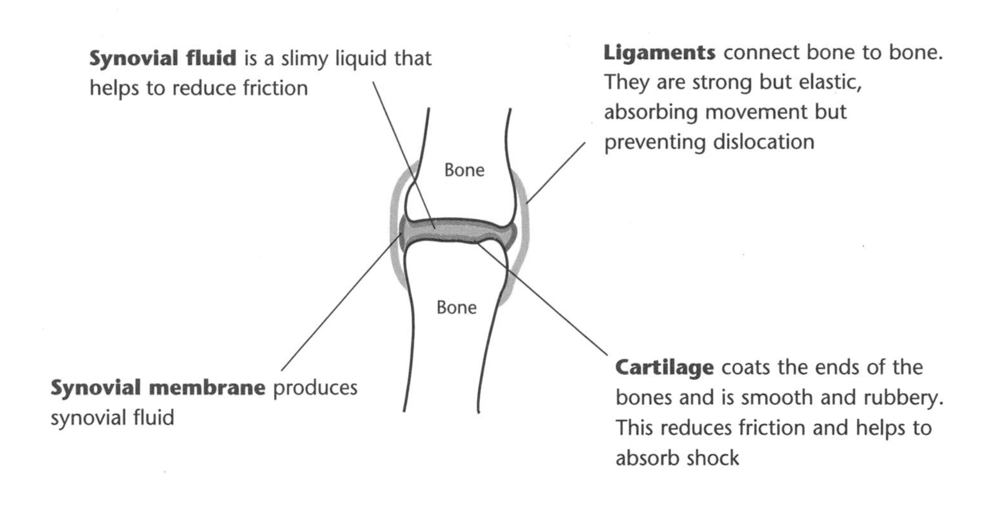

Fig. 10.2 A synovial joint.

Muscles and movement

Muscles are the main effectors in the body. They contain muscle fibres that can shorten and so make the muscle contract. In order for a muscle to contract, energy is needed from respiration. Muscles can contract and pull on bones, but they cannot actively expand or push.

Muscles have to be arranged in pairs called antagonistic pairs. When one contracts, the other relaxes and vice versa.

The biceps and triceps are antagonistic muscles in the human arm. The arm works like a lever, with the elbow as the pivot.

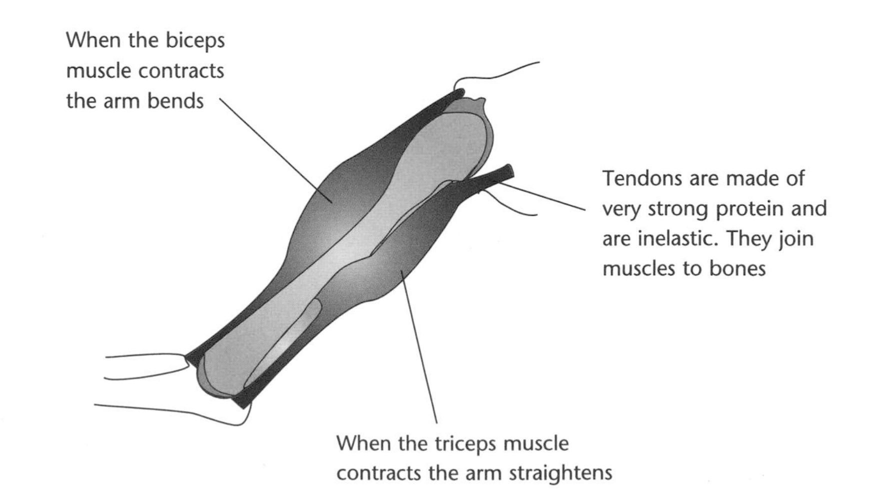

Muscles are connected to bones by **tendons**. Tendons are very strong and are not elastic.

Exercise and fitness

OCR A B7.7

Many people now visit a gym and regular exercise can improve a person's fitness. It can also improve their general health, helping to prevent diseases such as heart disease.

When a person decides to undertake a proper fitness programme, they should follow these steps:

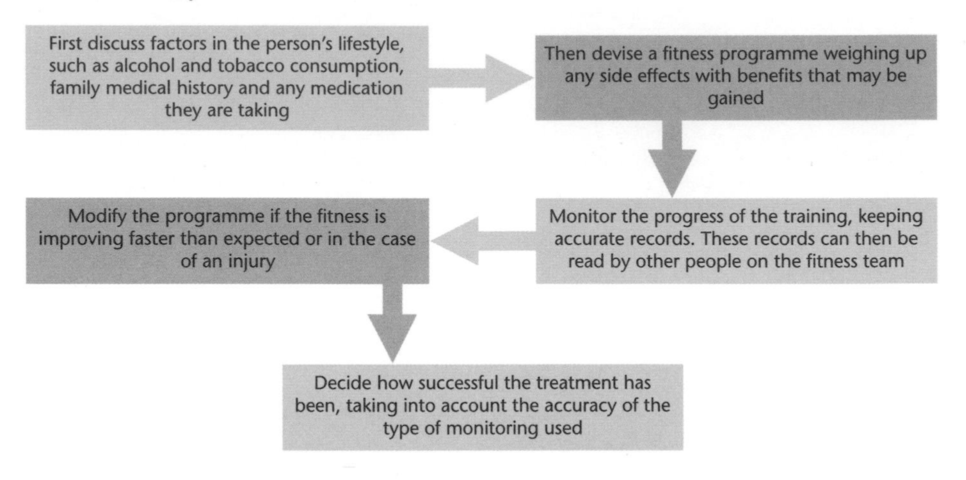

Common injuries

OCR A B7.7
OCR B B5a

Despite being strong, bones can be broken by a sharp knock. This is more likely to happen to elderly people because they often have **osteoporosis.** Osteoporosis makes their bones weaker.

There are three types of bone fractures:

- **simple** – a fracture of the bone without the skin being broken
- **compound** – a fracture of the bone where the skin has been broken
- **green stick** – here the bone is not fractured all the way through.

If a person has a fracture, they should not be moved as this could cause further damage (especially if it is a spinal injury).

Excessive exercise can cause damage to other parts of the body. These include:

- sprains
- dislocations
- torn ligaments
- damaged tendons.

The usual treatment for sprains is called RICE:

Rest

Ice

Compression

Elevation

These injuries may be treated by a physiotherapist who will devise a suitable set of exercises that should help the person recover.

10.2 The circulation

The heart and blood vessels

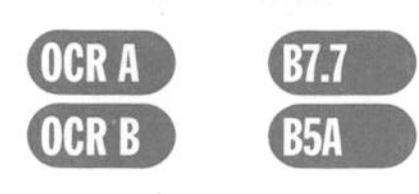

Our heart and blood vessels are organised in a double circulation. The advantages of this are discussed on pages 94–95. Our circulation has been investigated by many scientists throughout history:

Galen was doctor to five Roman emperors. He carried out dissections and showed that arteries carried blood, not air. He could not explain how the blood circulated.

Fig. 10.2 Galen.

Harvey lived from 1578 to 1627. He carried out experiments and showed that the blood flows from the heart in arteries and back in veins. He could not see capillaries, but guessed that they were there.

Fig. 10.3 Harvey.

The structure of the heart and blood vessels are shown on page 94.

> **KEY POINT**
>
> **The pattern of contraction of the different chambers of the heart is called the cardiac cycle:**
> - **first the atria fill up with blood**
> - **they then contract and force the blood into the ventricles**
> - **the ventricles then contract and force blood out into the arteries.**

Fig. 10.4 shows how the pressure changes in the left atrium, left ventricle and aorta as these events happen.

Remember, when a chamber contracts, the pressure of the blood will go up.

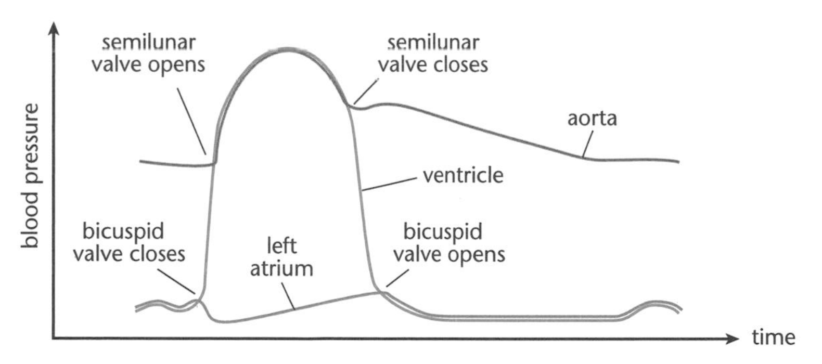

Fig. 10.4 Pressure changes during the cardiac cycle.

As the blood flows through the blood vessels, the pressure of the blood changes.

The blood in the arteries is at high pressure but when it reaches the veins, it is flowing smoothly with very low pressure.

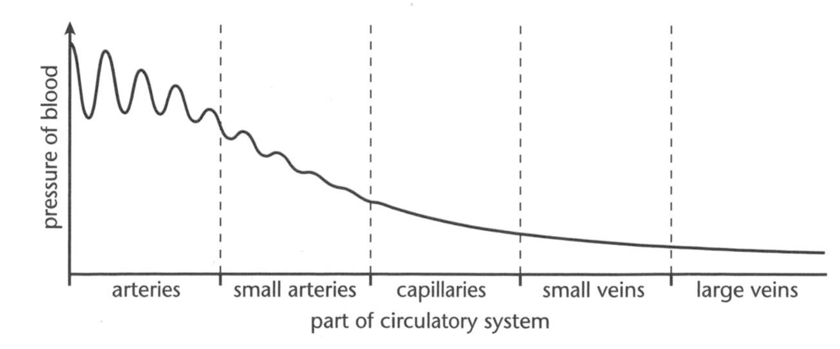

Fig. 10.5 Pressure changes through the circulatory system.

The heart is controlled by small groups of cells called the **pacemaker**. There are two areas called the SAN and the AVN. They produce a small electric current which spreads through the heart muscle making it contract. The rate of the pacemaker can be altered by hormones such as **adrenaline**.

The electric current that is produced by the pacemaker can be detected and studied using an **ECG** machine.

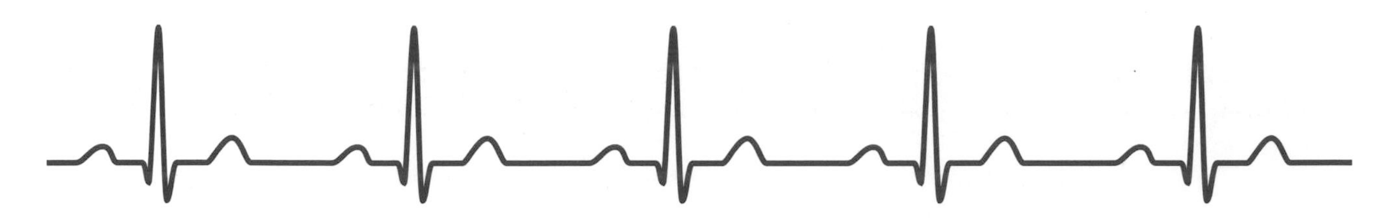

Fig. 10.6 An ECG brace.

Making repairs

Various parts of the circulation may need repair or replacement.

Blood

People have different blood groups. A person's blood group is controlled by their genes.

KEY POINT **One of the main sets of blood groups is the ABO system.**

The ABO system describes the chemical groups, or antigens, found on the red blood cells. It is determined by a single gene with three possible alleles: A, B or O. A and B are codominant and O is recessive to both.

When a person has a blood transfusion, it is important to make sure that the blood groups are matched. For example, if a person with blood group A is given blood of blood group B, antibodies in the person's blood will attack the antigens on the surface of the blood group B cells, and the blood will clot.

The heart

There are a number of problems that might occur with the heart. They can be treated in a number of different ways.

Condition or disease	Possible treatment
irregular heartbeat	use of an artificial pacemaker
hole in the heart	surgery to close the hole
damaged valves	replacement by artificial or animal valves
coronary heart disease	bypass surgery, use of a heart assist device or heart transplants

Some of these conditions are inherited, but many are related to the way a person lives their life. These are called **lifestyle factors** e.g. diet, alcohol intake, drugs, stress and smoking.

10.3 Respiratory systems

Gaseous exchange and breathing

Different types of organisms have different systems for obtaining oxygen and losing carbon dioxide.

KEY POINT **The moving of these gases between the organism and the environment is called gaseous exchange.**

Small organisms have a large surface area to volume ratio.

In small organisms such as worms, gaseous exchange occurs over the whole body surface.

Fish have **gills** for gaseous exchange. The filaments have a rich blood supply and a large surface area to take in oxygen from the water:

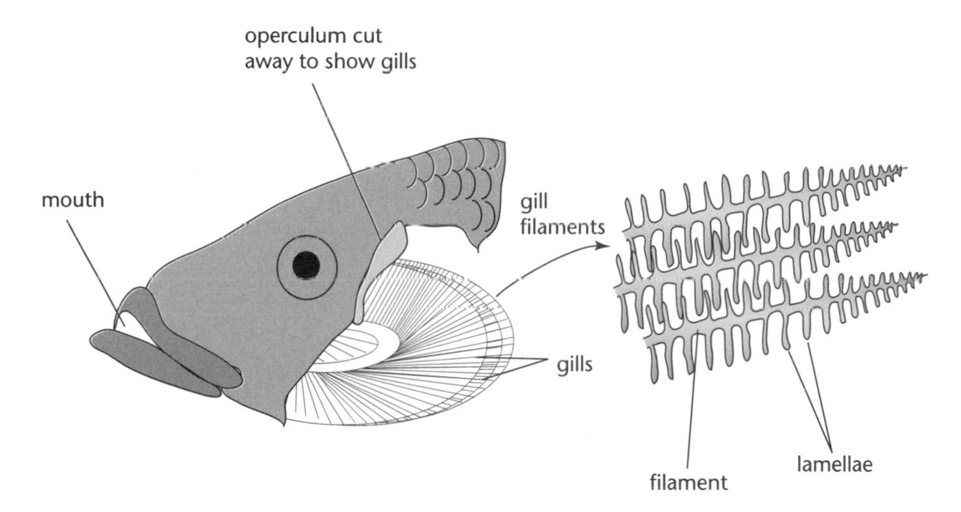

Fig. 10.7 Gill structure.

The presence of gills on fish means that they must live in water. Amphibians, such as frogs, have lungs but also obtain oxygen through their skin. Their skin needs to be moist so while they are able to live on land, they can only survive in damp places.

Humans have **lungs** for gaseous exchange. They have millions of tiny air sacs called **alveoli.** Humans have various adaptations for maximum gaseous exchange:

- the millions of alveoli provide a surface area of about 90 m^2
- the many blood vessels provide a rich blood supply
- the alveoli have a thin film of moisture, so that the gases can dissolve
- the blood and air are separated by only two layers of cells.

To speed up gaseous exchange, the body uses the intercostal muscles and diaphragm to draw air into and out of the lungs. This is called **breathing**. The volume of air that is drawn in and out can be measured on a machine called a **spirometer.**

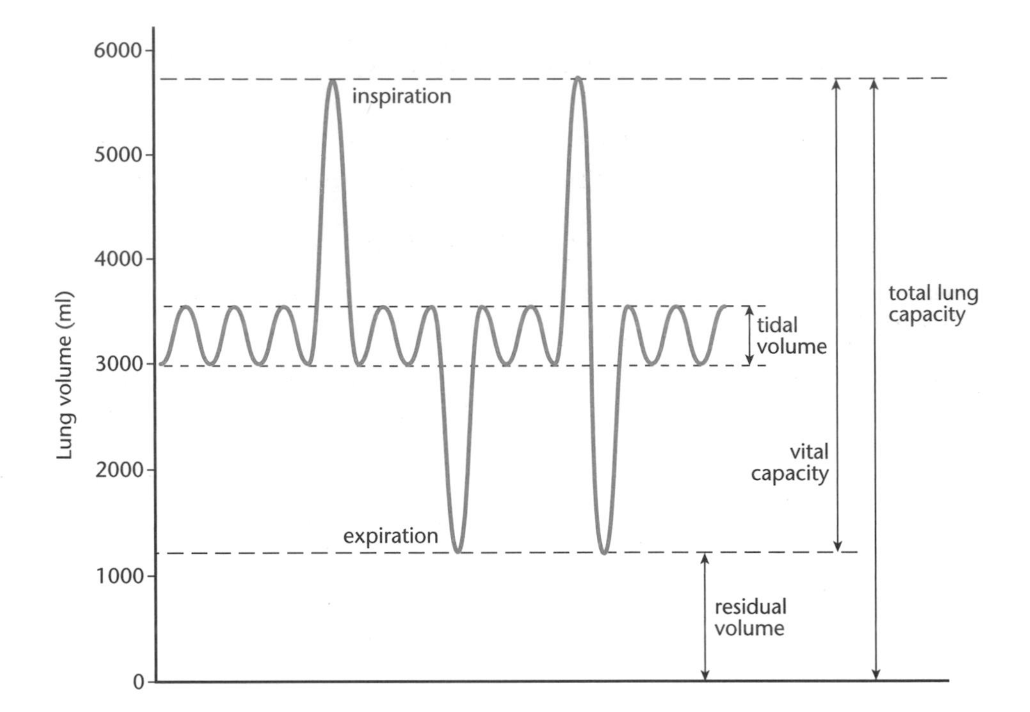

Fig. 10.8 A spirometer trace.

Total lung capacity is the maximum volume of air that the lungs can hold.

Vital capacity is the maximum air that the lungs can pass in or out per breath.

Residual volume is the air that is left in the lungs after a person has breathed out deeply.

Respiration and exercise

OCR A B7.5
AQA B3.13.3

The oxygen that is used for gaseous exchange is used in **aerobic respiration**. Page 11 contains the equation for aerobic respiration and describes changes occurring in the circulation and lungs during exercise.

Changes also occur in the muscles:

- stored glycogen is changed to glucose for respiration
- blood vessels supplying the muscle dilate to give the muscle more blood.

Page 11 also explains what happens during exercise if the muscles cannot be supplied with enough oxygen. This results in **anaerobic respiration** and the production of lactic acid.

The use of anaerobic respiration to make alcohol is covered on page 115.

Despite the drawbacks of anaerobic respiration, it is useful for a number of reasons:

- it is used by animals for short bursts of energy
- some parts of plants may use it, e.g. germinating seeds
- some microbes might use it, e.g. yeast.

Both aerobic and anaerobic respiration release energy that is then trapped in a molecule called **ATP.** This is often called the **energy currency** of living organisms because it can be used to power all the processes that require energy.

Lung problems

OCR B B5d

The lungs are easily infected because they are a 'dead end'. Microbes and particles can easily collect there.

Some of these types of problems are:

Some of the actions of smoking on the lungs are described on page 32.

- **industrial diseases** such as asbestosis where small particles are breathed in and damage the lungs
- **genetic conditions** such as cystic fibrosis where too much mucus is made
- **lifestyle factors** such as smoking.

The respiratory system tries to protect itself from disease by producing mucus and the action of cilia. These are shown on page 27.

It is possible to have a transplant if the lungs are too badly damaged. This may involve a heart and lung transplant. During some operations, the functions of the heart and lungs are taken over by a device outside the body called a **heart-lung machine**.

10.4 Removing waste

Different wastes

OCR B B5e

The body produces different types of waste as a result of its metabolism. Many of these wastes are toxic and so must be removed from the body.

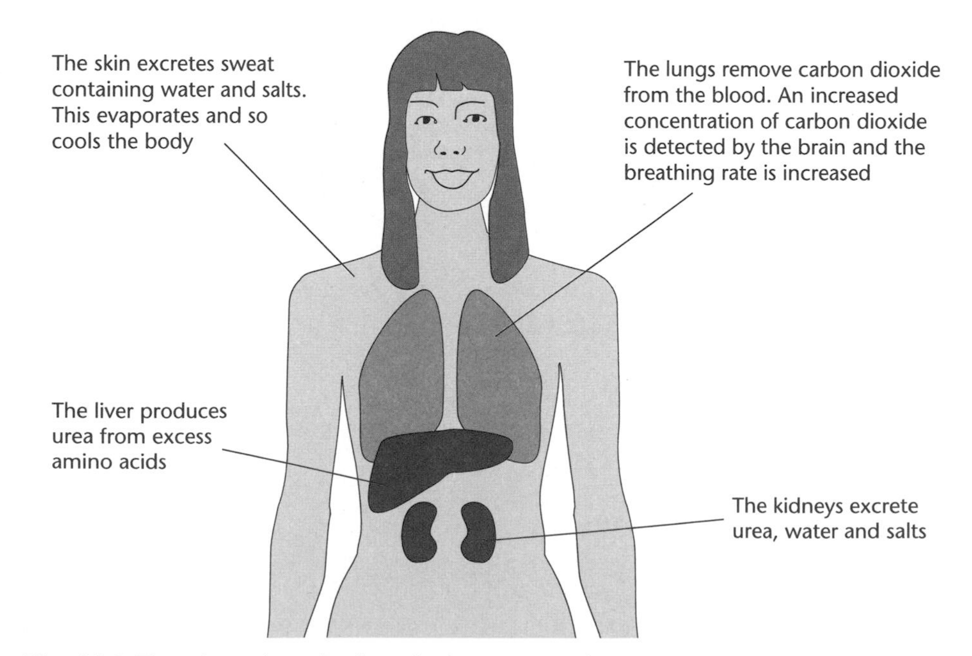

Fig. 10.9 The sites of production of excretory products.

How the kidneys work

OCR B B5e
AQA B3.13.4

The kidneys produce urine by the following process:

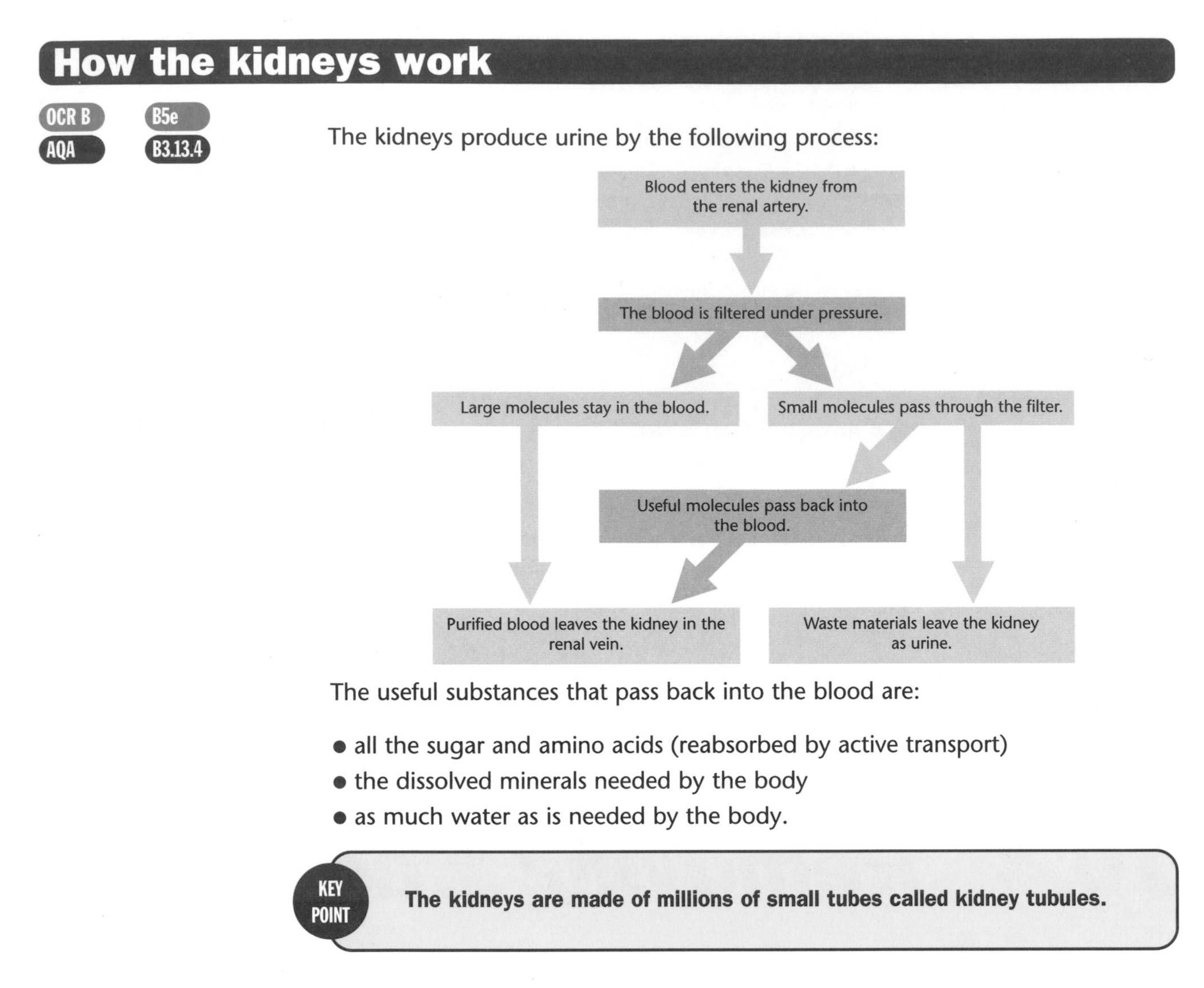

The useful substances that pass back into the blood are:

- all the sugar and amino acids (reabsorbed by active transport)
- the dissolved minerals needed by the body
- as much water as is needed by the body.

KEY POINT **The kidneys are made of millions of small tubes called kidney tubules.**

Different parts of these tubules do different jobs in the production of urine.

Only OCR B candidates need to know the structure of the kidney.

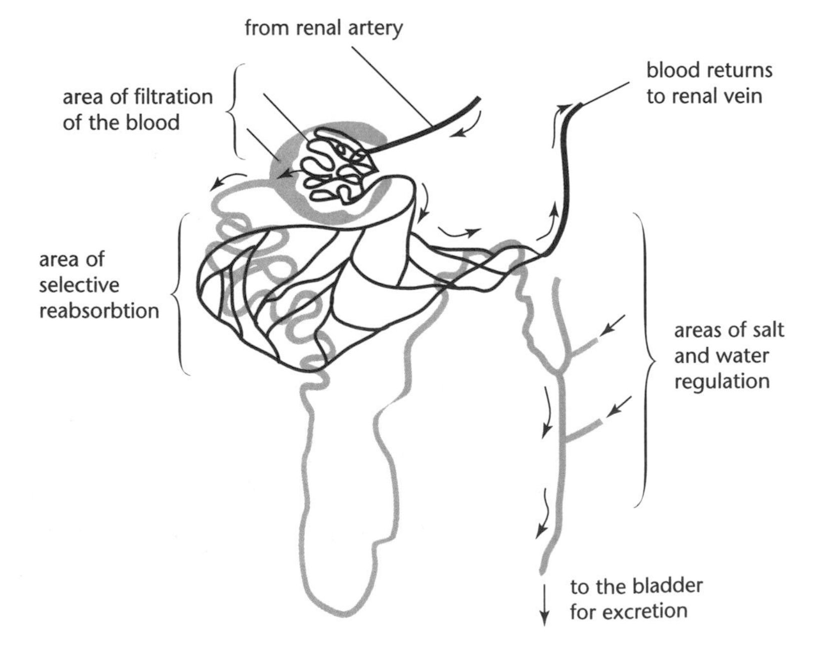

Fig. 10.10 Structure of a kidney tubule.

The job of the kidneys in controlling the water balance of the body involves the hormone ADH. This is described on page 94.

Treating kidney disease

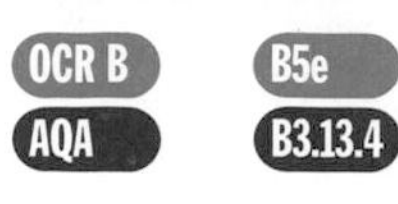

People may have kidney failure for a number of reasons. A person can survive if half of their kidney tubules are still working, but if the situation worsens, there are two options:

Kidney dialysis

Kidney dialysis involves linking the person up to a dialysis machine. This takes over the job of the kidneys and removes waste substances from the blood.

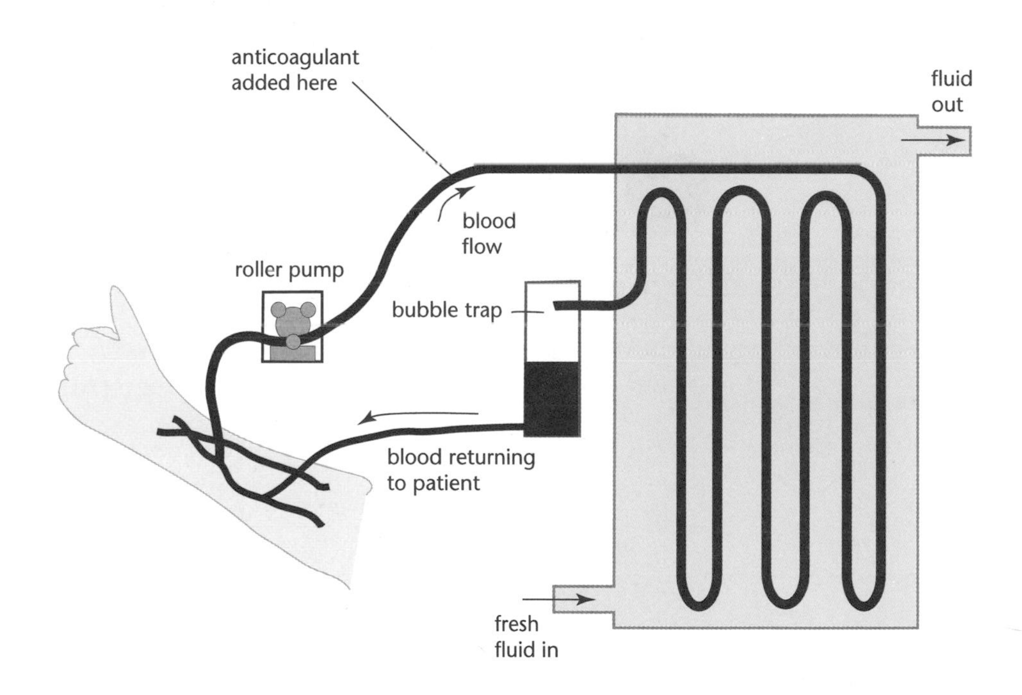

Fig. 10.11 A dialysis machine.

The blood is removed from a vein in the arm. It then passes through a long coiled tube made of partially permeable cellophane. The fluid surrounding the tube contains water, salts, glucose and amino acids but no waste materials, such as urea. These waste materials therefore diffuse out of the blood into the fluid.

Transplants

As a person can survive with one kidney, it is possible for a person to donate one kidney to be transplanted into another person. Other transplants may come from dead donors.

The main problem with transplants is the person's immune system rejecting the transplanted kidney.

This is avoided by taking certain precautions:

- making sure that the donor has a similar 'tissue type' to the patient
- treating the patient with drugs to make their immune system less effective.

Some of the issues involving organ donation are covered on pages 99–100.

10.5 Reproduction and growth

Fertility and development

 OCR B B5f, h, a

Fertilisation is the result of fusion between a male gamete (sperm) and a female gamete (egg).

Make sure you know the structure of the male and female reproductive systems.

In males after puberty, sperm production is a continual process, but egg production in the female follows a regular cycle. This also involves changes in the uterus. These changes are called the **menstrual cycle**.

The menstrual cycle is controlled by four main hormones.

Hormone	Possible treatment
FSH	Stimulates the growth of a follicle containing an egg
oestrogen	Repairs the lining of the uterus
LH	Stimulates the release of an egg (ovulation) and then stimulates progesterone production
progesterone	Further repairs the lining of the uterus and stops it breaking down

Further details of these hormones and the menstrual cycle are on pages 19–21.

Infertility

About one in seven of all couples have problems having children. There are various treatments for **infertility** and each has different issues involved.

Treatments	Possible issues
artificial insemination	Can be used to concentrate sperm if sperm count is low, or use of sperm from a donor
use of FSH	Can stimulate egg production but may lead to multiple births
IVF	Can bypass blocked oviducts but there may be disputes over any frozen embryos
egg donation	A woman can receive donated eggs if she cannot produce her own but the child will not receive any of the woman's genes
surrogacy	Another woman may be prepared to give birth to a baby for another couple, but may be reluctant to give up the baby once it is born
ovary transplants	A woman can receive ovary transplants, but the eggs will not be genetically hers if the ovary tissue is from another woman

Growth and development

As soon as a woman becomes pregnant, the growth and development of the baby is monitored:

- Tests can be performed on the pregnant woman such as **amniocentesis.** Amniocentesis involves taking some liquid from around the baby and testing the cells for conditions such as Down's syndrome. There is a slight risk of miscarriage with the procedure. Also, the parents may have to face a difficult decision about whether to have the baby or not if the tests show abnormalities.
- Tests are carried out once the baby is born, such as measuring their length, mass and head size. These are compared to average growth charts like the graphs shown on page 73.

Growth in animals is largely due to new cells being made by mitosis.

Growth can be altered by various factors:

The growth hormone causes growth in long bones. Dwarfs produce too little of the hormone, and giants produce too much.

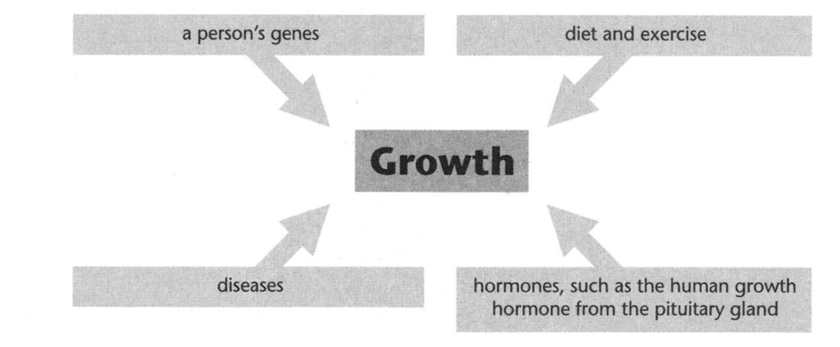

As a person grows, the cartilage in their skeleton gradually gets replaced by bone (ossification). The amount of cartilage present can be used to decide how old a person is.

Can genetic modification beat malaria?

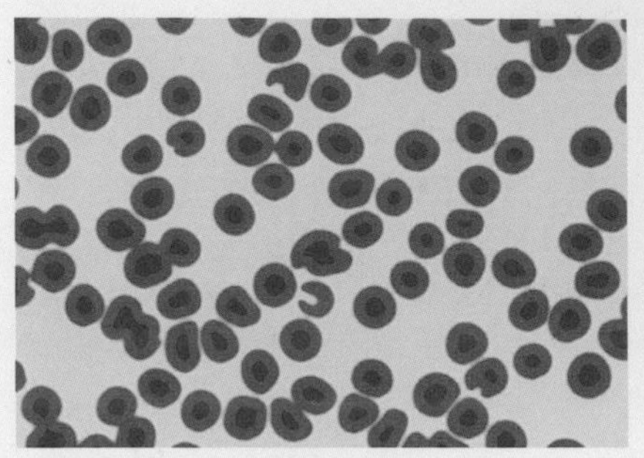

Plasmodium, the malaria parasite inside blood cells.

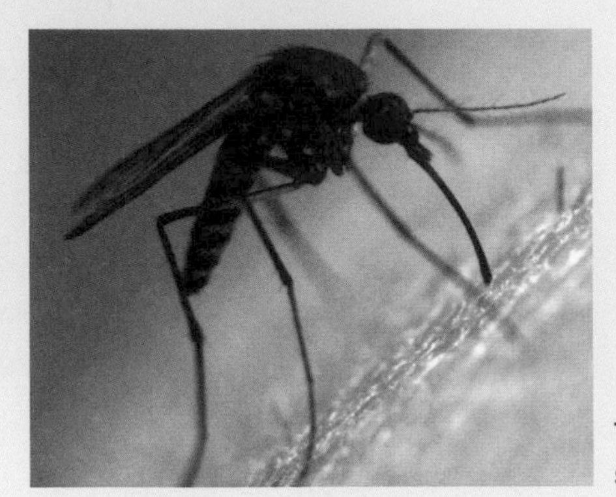

The mosquito that spreads malaria.

Malaria kills more people in the world than any other infection except HIV/AIDS. It infects between 300 million and 500 million people each year and kills more than one million.

It is not caused by a virus or a bacterium, but single-celled protozoan called *Plasmodium.* This parasite is spread from person to person by female mosquitoes when they suck blood. Fortunately for people in Britain, the mosquito does not live here, it lives mainly in Africa and South America. With global warming, however, who knows where it may be able to live in the future?

Protecting people from malaria has proved to be very difficult.

Some people have natural resistance to the malaria parasite, but this is because they have a genetic condition called sickle cell anaemia. Carriers of this condition are protected from malaria, but unfortunately, individuals who have two copies of the gene have sickle cell anaemia, which can be lethal.

There are drugs that can treat malaria, but they are expensive and no vaccine is yet available. Perhaps genetic modification may help?

Genetically modified mosquitoes

The malaria parasite usually uses the mosquito as a carrier or vector. It does not kill mosquitoes, but it does make their breeding less efficient. Scientists in America have produced genetically engineered mosquitoes that will not carry the *Plasmodium* parasite. They think that releasing them into the wild might cause them to replace the normal mosquito because of this advantage. Malaria cannot then be passed on.

Genetically modified parasite

Scientists in Germany are hoping that they might have produced a vaccine that can be used to prevent malaria. They have produced a genetically modified version of the malaria parasite, *Plasmodium*. This GM version of *Plasmodium* attacks rats, but is harmless. They have found that injecting the rats with this parasite prevents them from being attacked by the deadly version of *Plasmodium*. The GM version is acting as a vaccine. The scientists hope that they might use this approach to make a human vaccine.

HOW SCIENCE WORKS

Genetically modified bacteria

A drug called artemisinin is quite effective at treating malaria. This drug is made from a chemical extracted from a plant called sweet wormwood. The plant has been used for thousands of years in China to treat malaria. Now scientists have found that it is 97% effective in curing malaria. The problem is that the plant takes six months to grow and months to extract the chemical. A dose of the drug costs £1.35 per person. Scientists have now put genes from wormwood into bacteria and the GM bacteria can produce the chemical needed. They hope that the cost will come down to 14p per dose.

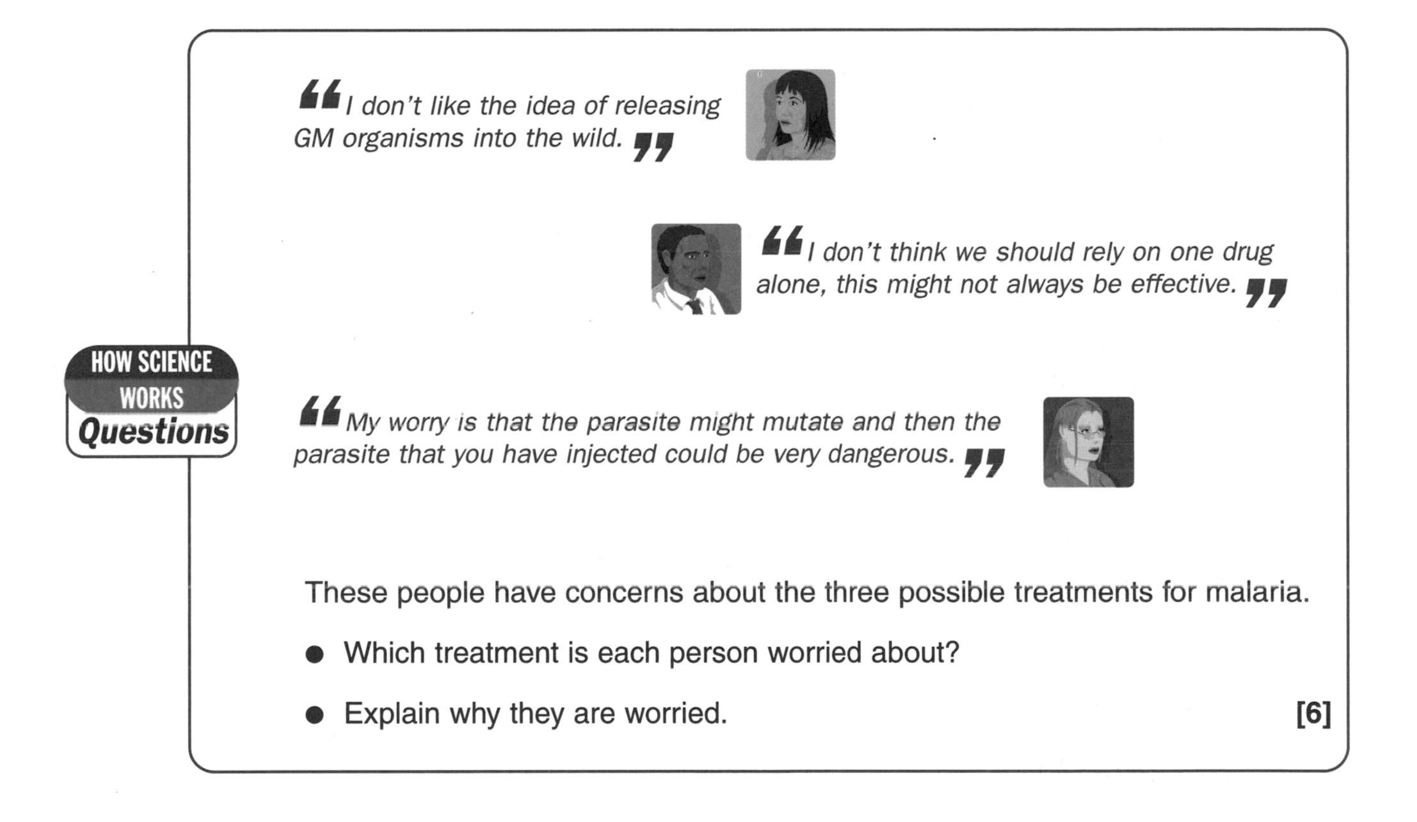

These people have concerns about the three possible treatments for malaria.

- Which treatment is each person worried about?
- Explain why they are worried. **[6]**

Exam practice questions

1. The following terms are involved in human physiology.
Match words **A, B, C**, and **D**, with the numbers **1–4** in the sentences.

A amniocentesis
B dialysis
C pacemaker
D osteoporosis

A condition that weakens the bones is called ____1____.
An irregular heart beat may be treated by an electrical ____2____.
The function of the kidneys may be replaced by ____3____.
An unborn baby may have its cells investigated using ____4____. **[4]**

2. Put a tick (✓) or a cross (✗) in each empty box to show how the kidney deals with glucose, protein and urea.

	Glucose	Protein	Urea
Present in the blood reaching the kidney			
Passes out of the blood in the filter unit			
Reabsorbed back into the blood from the kidney tubule			
Usually present in the urine			

[4]

3. The boxes contain some names of parts of the body and descriptions.
Draw straight lines to join each **Part** to the correct **Description**.

Part	**Description**
bone	Living tissue containing cells and calcium salts
cartilage	Elastic structure that holds joints together
ligaments	Inelastic structure that joins muscle to bone
tendons	Shiny substance that reduces friction in joints

[3]

Exam practice questions

4. Look at Fig. 10.4 and Fig. 10.5 on page 127 and use the information to answer the following questions.
   (a) What is the highest pressure that the blood reaches in the
   (i) left ventricle
   (ii) left atrium? **[2]**
   (b) Explain why there is such a difference between these two pressures. **[2]**
   (c) How do you think the pressure would differ in the **right** ventricle compared to the left? Explain your answer. **[2]**
   (d) Fig. 10.5 shows why veins have valves. Why is this? **[2]**

5. Jack has asthma.
   This means that something causes the small tubes in his lungs to narrow. It makes it harder to draw air into the small air sacs where gaseous exchange occurs.
   (a) **(i)** What is the name of these air sacs? **[1]**
   (ii) Suggest one factor that might make the small tubes narrow. **[1]**
   (b) Jack decides to measure his breathing using a spirometer.
   He makes a trace when he is breathing normally and when he has an asthma attack.

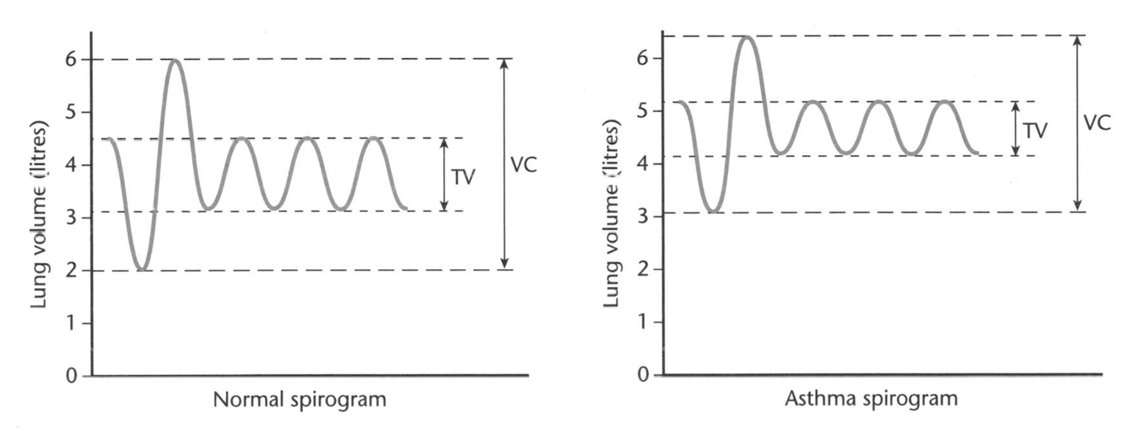

   (i) What do the letters **VC** and **TV** stand for? **[2]**
   (ii) Describe the differences in Jack's breathing when he has an asthma attack compared to normal breathing. **[4]**

6. Look back at the How Science Works article about malaria.
   (a) Which of the following lines in the table best describe the organisms mentioned in the article? Put a tick in the correct box.

Pathogen	Vector	Host	
Plasmodium	mosquito	human	
human	*Plasmodium*	mosquito	
mosquito	human	*Plasmodium*	

   (b) Why is it so important to produce an inexpensive treatment for malaria? **[2]**
   (c) Explain why bacteria can produce the chemical needed to make the drug at a cheaper cost than plants. **[2]**

Chapter 11 Animal behaviour

The following topics are covered in this chapter:

- *Instinctive or learned behaviour?*
- *Feeding behaviour*
- *Human behaviour*
- *Social behaviour and communication*
- *Reproduction*

11.1 Instinctive or learned behaviour?

Instinctive behaviour

EDEXCEL B3.2

KEY POINT: **When animals are born, they have certain inbuilt types of behaviour. These are called instincts.**

Instincts are much more complicated actions than reflexes.

Instinctive behaviour is controlled by the genes and is inherited from the animal's parents. Instinctive behaviour is important so that the young animal has certain skills needed to survive before it has the chance to learn.

Examples include ducklings being able to swim straight after they have hatched from eggs.

Learned behaviour

EDEXCEL B3.2

As soon as a young animal is born it starts to change its behaviour. This is because it starts to learn.

KEY POINT: **Learning is a change in behaviour caused by experiences.**

One way that animals learn is by **habituation**. An animal may be frightened by a particular stimulus such as a load noise. However, if the stimulus is repeated and the animal is not harmed then the animal learns to ignore the stimulus. It has been habituated. This idea is often used in the training of animals such as horses.

Animals can also learn by **conditioning**. Instead of learning to ignore a stimulus they will associate one stimulus with another. Examples of this are on page 97.

This type of learning is also used in training animals for specific jobs.

11.2 Social behaviour and communication

Communication

EDEXCEL B3.2

Many animals live together in various types of social groups. This makes it necessary for them to communicate with each other. There are a number of ways that animals can do this:

- sounds
- signals
- chemicals such as pheromones.

In mammals such as chimpanzees, gorillas and humans, a lot of information is exchanged by **body language**. This includes body posture and facial expressions.

Fig. 11.1 Different facial expressions in chimps.

Fig. 11.1 shows some facial expressions in chimps, but the same expressions may mean something completely different to another type of animal. They are **species-specific**.

Human communication

EDEXCEL B3.2

Unlike other animals, humans can communicate using true language.

True language uses symbols such as written words or spoken words to pass on messages. These messages can be complicated abstract ideas such as:

- memories of past events
- feelings and emotions.

Humans spend a long time as children, dependent on their parents. This gives them the chance to learn from their parents. It allows cultural ideas to be passed on.

Humans are also more conscious of the outcomes of their actions than other animals. Up until recently it was thought that only humans, apes and dolphins can recognise themselves in a mirror. This is taken as a sign that they are **self-aware**.

11.3 Feeding behaviour

Herbivore and carnivore behaviour

EDEXCEL B3.2

Animals have different feeding behaviour depending on the type of food that they eat.

KEY POINT

Herbivores only eat plant material and so their behaviour is adapted in a number of ways.

Carnivores only eat animals and so their behaviour is adapted in a number of ways.

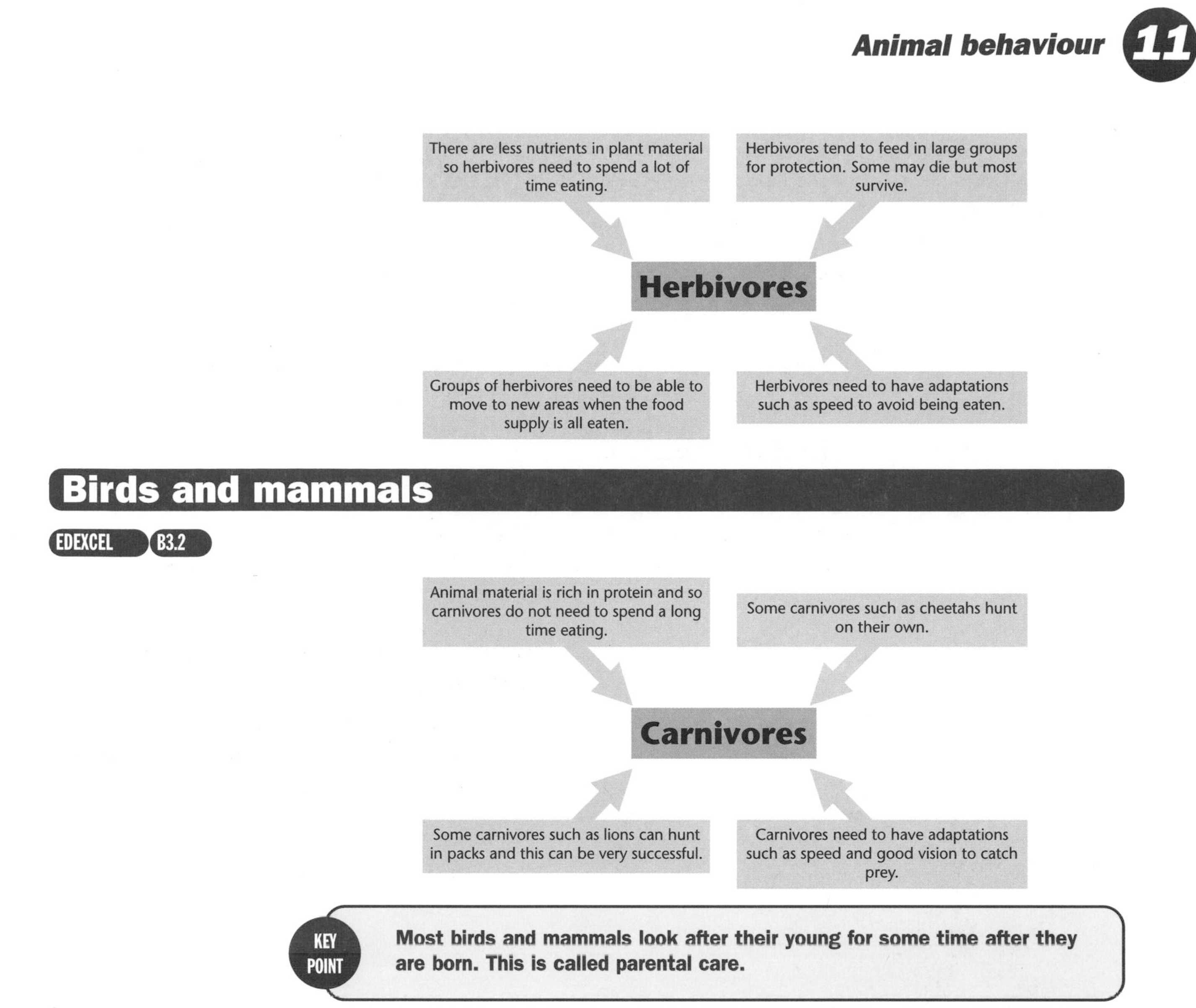

Birds and mammals

EDEXCEL B3.2

Animal material is rich in protein and so carnivores do not need to spend a long time eating.

Some carnivores such as cheetahs hunt on their own.

Carnivores

Some carnivores such as lions can hunt in packs and this can be very successful.

Carnivores need to have adaptations such as speed and good vision to catch prey.

KEY POINT **Most birds and mammals look after their young for some time after they are born. This is called parental care.**

This means that these animals have developed special feeding behaviour.

Baby birds will call to their parents and often gape with their mouth wide open. Often the inside of the baby's mouth is coloured to attract the parent bird.

Fig. 11.2 Young birds showing feeding behaviour.

Mammals feed their young on breast milk. The young are born with an instinct to suck on the breast. The process of breast feeding is also important in building an emotional bond between the mother and the baby.

Fig. 11.3 Breast feeding builds an emotional bond.

It use to be thought that only humans used tools to get food. However, in the 1960s chimpanzees in the wild were seen stripping leaves off branches to make sticks for fishing termites out of their nests. Capuchin monkeys have been seen to use rocks to smash open palm nuts.

Even birds have uses for tools. Woodpecker finches can use a twig to pull grubs and insects from holes in trees. Another bird, the Egyptian vulture, throws rocks at ostrich eggs to break them open so they can eat them.

11.4 Reproduction

Finding a mate

EDEXCEL B3.2

Sexual reproduction involves individuals choosing who to mate with. This is particularly important for females as they often use energy and food reserves producing the young.

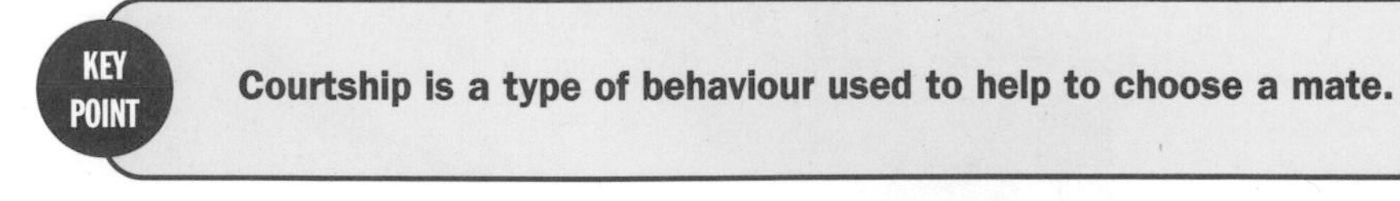

KEY POINT

Courtship is a type of behaviour used to help to choose a mate.

In many types of animals, males show courtship behaviour to try and persuade females that they have good genes and can provide for the offspring.

Once a mate is chosen, there are a number of different types of relationship:

- very few animals mate with the same partner for life. Albatrosses can live to be 80–85 years old and they mate for life
- some animals live in groups with a dominant male that mates with all the females

- other animals may have different mates each year or even in the same breeding season.

Once the offspring is born, animals may or may not show parental care. This has advantages and disadvantages:

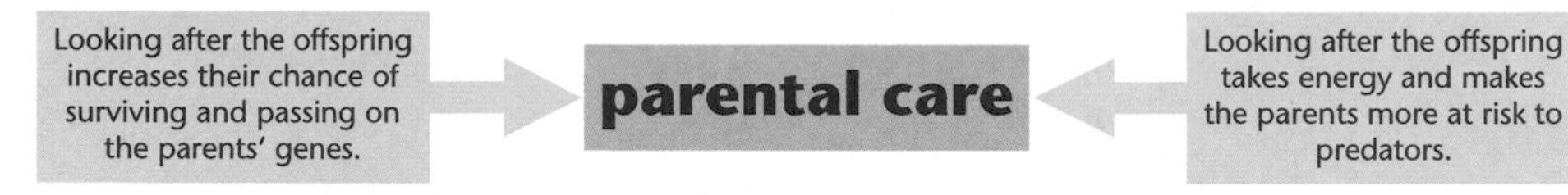

11.5 Human behaviour

Our relationships with other animals

EDEXCEL B3.2

Humans are great apes along with gorillas, orangutans, chimpanzees and bonobos (pigmy chimps).

Great apes are not monkeys. Apes are larger, spend more time upright, and depend more on their eyes than on their noses and do not have tails. They are more intelligent than monkeys.

Humans are thought to have developed from small groups of animals called **hunter-gatherers**. This means that they lived by collecting food from the wild.

Humans have now evolved away from being hunter-gatherers. They have developed the ability to change their environment and use other animals in many ways:

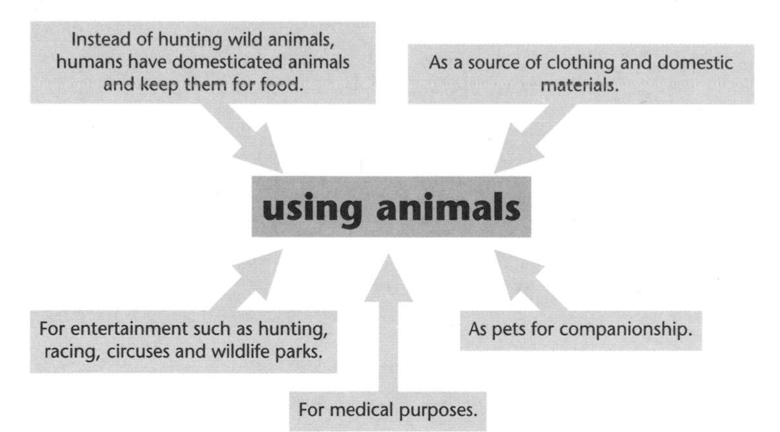

Many people argue about the way humans use animals. Some think that animals should have the same rights as people, whilst others disagree.

Sometimes it is possible to assume that animal behaviour is very similar to human behaviour.

This is called anthropomorphism.

Whilst there are similarities, scientists have to be careful not to assume some behaviour is the same as human behaviour.

HOW SCIENCE WORKS

What makes us human?

People have always argued about what makes humans different from other animals. Is there one large difference or is it a number of smaller differences? Several characteristics are often mentioned as main differences between us and other animals. These include:

- self-awareness
- language
- culture.

But do other animals share any of these?

Elephant self-awareness

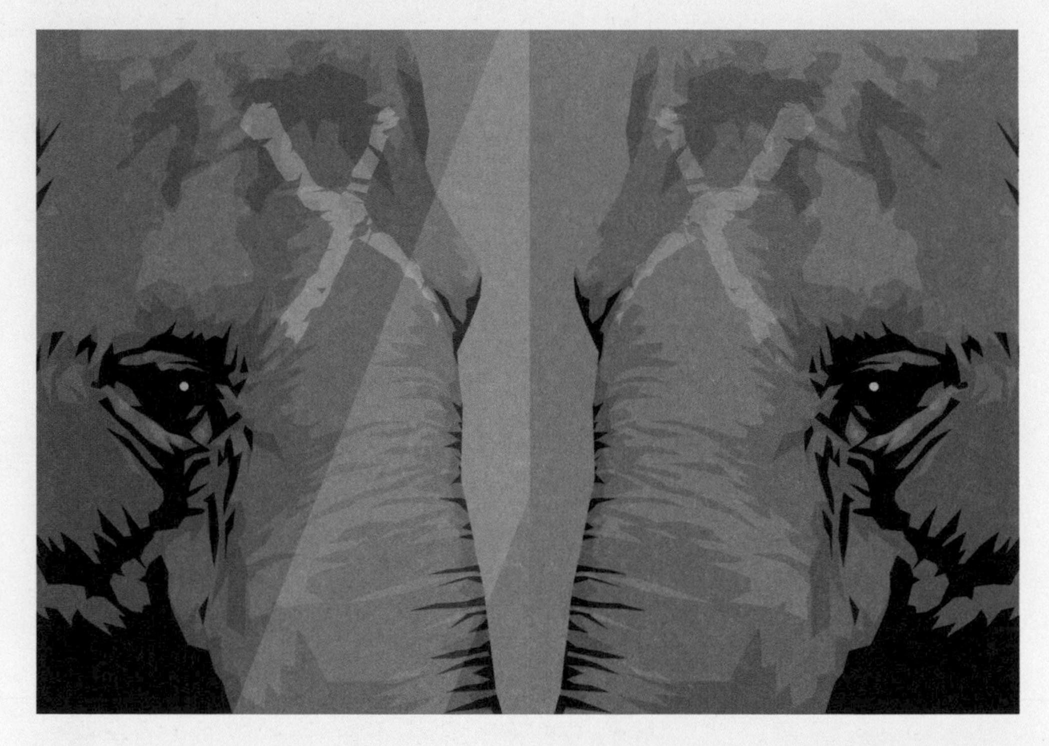

Scientists now think that elephants may be self-aware. The scientists investigated three adult female Asian elephants. They put a mirror eight feet wide by eight feet high in their yard at the Bronx Zoo. Remarkably, the elephants did not seem to mistake their reflections for other elephants and try to greet them. Finally, once the animals recognised the reflections as their own, they used mirrors to investigate their own bodies. The elephants stuck their trunks into their mouths in front of the mirror, and one used her trunk to pull her ear slowly toward the mirror. One elephant, passed the final test of repeatedly touching an X painted on her head, a place she could not see without a mirror. This is an important test and more than half of chimpanzees often fail this test.

Humans can usually recognise their reflection when they are eighteen months old.

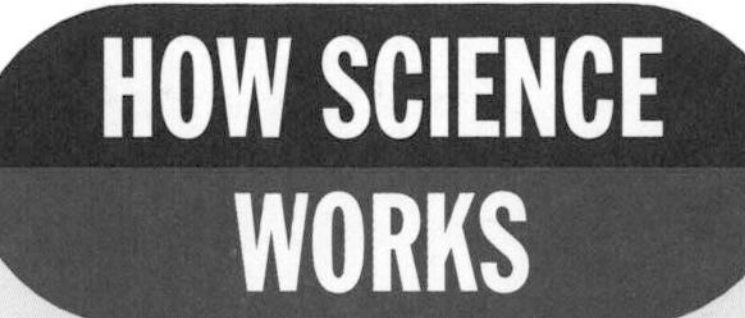

Do apes use language?

Scientists often disagree as to whether apes can use language. Some scientists say they cannot because the ape brain is not capable of processing language. Others say that they can use language, but cannot speak because they are physically unable to make the correct sounds.

There have been many attempts to teach apes to talk, but all have failed. In the early 1970s, a chimpanzee named Washoe was taught to communicate in Sign Language. Washoe learned 132 different words and became the first non-human to learn symbols and use them to communicate. Four other chimps have also been taught to sign. These five chimps now sign not only to the humans, but also to each other to communicate.

Many scientists, however, still believe that apes have no real grasp of human language, but are just copying humans. They say that while apes may understand symbols or words, they do not understand how words are put together to form a complete idea.

Do monkeys have culture?

Culture is often described as the ability to invent new behaviour that is adopted by the population and is passed down through the generations. In 1948, scientists visiting a Japanese island encountered a troop of wild monkeys. To observe the monkeys, they tempted them out of the forest with sweet potatoes and wheat. In 1953, a young female named Imo was the first to wash the sweet potatoes. She passed this behaviour to her mother and it slowly began to spread. Ten years later, potato washing had become a fixed behaviour in the group. Most newborns monkeys picked up the skill quickly. Imo then developed a method for sorting wheat from sand. She dropped a mixture of wheat and sand in water allowing the wheat to float and the sand to sink. Within several years many of the younger monkeys did this.

HOW SCIENCE WORKS Questions

These observations of animal behaviour raise some interesting questions:

1. Do you think that there is one main difference between humans and other animals? **[2]**
2. Why do scientists have to be careful when trying to draw conclusions from looking at animal behaviour? **[2]**

Exam practice questions

1. The following words are used in describing behaviour in apes.
Match words **A, B, C**, and **D**, with the numbers **1–4** in the sentences.

A pheromones
B learning
C body language
D culture

Apes have the ability to communicate using expressions, this is an example of ____1____.
They can also communicate using chemicals called ____2____.
It has been shown that apes can change their behaviour as a result of experiences.
This is called ____3____.
If this change in behaviour is passed on to other generations, it is an example of ____4____. **[4]**

2. The boxes contain some types of behaviour and some examples.
Draw straight lines to join each **Type of behaviour** to the correct **Example**.

Type of behaviour	Example
courtship	Police dogs have been trained not to be frightened of loud noises
habituation	Male peacocks often display their tail feathers
instinct	Car drivers usually brake when they see a red traffic light without thinking about it
conditioning	Baby chicks will follow the first moving object that they see

[3]

3. An experiment was set up to investigate the attacks of a predator bird, a goshawk, on pigeons.
The attack success of the goshawk against different-sized flocks of pigeons was studied. The results are shown on the graph.

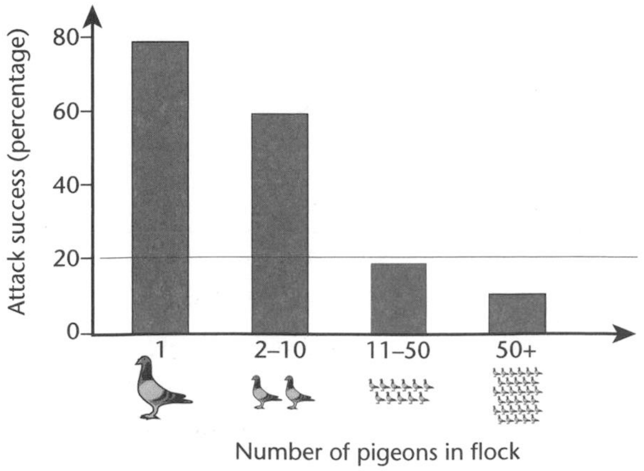

Attack success (percentage)
80
60
40
20
0
1
2–10
11–50
50+
Number of pigeons in flock

Exam practice questions

(a) Describe what the results of the experiment show. **[2]**
(b) Suggest a possible explanation for the results. **[2]**
(c) Suggest a possible **disadvantage** for prey animals living in large groups. **[1]**

4. Grebes are birds that show courtship behaviour. This is shown in the diagrams.

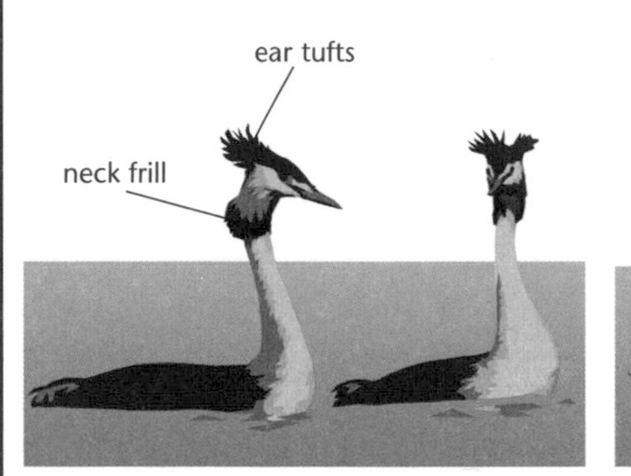

Before courtship | Courtship begins | The grebes then dance and give each other waterweed

(a) Why is courtship behaviour important to birds? **[2]**
(b) Write down one difference between the grebes once they have started their courtship behaviour compared to before. **[1]**
(c) Suggest why the grebes give each other water weed during their dance. **[1]**

5. The diagram shows an apparatus used by the scientist Ivan Pavlov in 1905.
He operated on the dog to insert a tube into its salivary gland to collect saliva.

Pavlov gave the dog meat powder just after a ticking noise. The dog produced saliva.

(a) The experiment was repeated many times and then the dog was exposed to the ticking noise without any meat powder.
Suggest what happened. **[1]**
(b) Explain your answer to **(a)**. **[2]**
(c) Some people may now be rather unhappy about this type of experiment.
Write about the arguments for and against this type of experiment. **[4]**

Answers

Chapter 1

How Science Works

1. two from: It is unlikely that a large number of embryos will all survive if left to develop; Removing some may allow the others to survive; This would result in the destruction of some of the embryos; Some people would consider this murder.
2. two from: less fit to give birth; less fit to bring up children; more likely to die before the children have grown up; may find it harder to relate to the children as they grow up.

Exam practice questions

1. **D**
2. **A**
3. **A**
4. **1** Growth **2** amino acids **3** iron **4** digested **5** absorption
5. **(a)** **(i)** Pancreas **(ii)** liver
   (b) Needed for respiration by cells; if it becomes too high it passes out in the urine
   (c) Five from: glucose diffuses into the blood stream/in the small intestine/rise in blood sugar level causes release of insulin/from the pancreas/causes liver to store more glucose/as glycogen
6. **(a)** Hormone A: oestrogen; first half
   Hormone B: progesterone; second half
   (b) Ovaries
   (c) Higher in the second half of the cycle; small fluctuations
   (d) **(i)** Largest increase in temperature around time of ovulation; intercourse then is more likely to lead to pregnancy
   (ii) Avoid intercourse around the time of the temperature increase
   (e) Women do not all ovulate regularly/may have intercourse the day before a temperature increase/does not give any protection against STDs

Chapter 2

How Science Works

1. to see if they work; to see if they are safe
2. so that a person does not know if they have taken a drug or not; to eliminate psychological effects;
3. For: animals were suffering unnecessarily; they do not always react in the same way as people
   Against: This is only one case whereas thousands of drugs have been tested; Testing on animals may have saved thousands of lives

Exam practice questions

1. **B**
2. **A**
3. **D**
4. Pathogen; lysozyme; acid; toxins
5. **(a)** Measles can cause babies to die/mumps can cause deafness in young children
   (b) Because if they catch rubella their babies may become brain damaged
   (c) Three from: contains a weakened or dead pathogen/stimulates the production of antibodies/from white blood cells/memory cells are made/if the live pathogen invades it can be killed rapidly
   (d) It often contains a live but weakened pathogen; people get a mild form of the disease
6. **(a)** Nicotine
   (b) Cannot do without it; lack of the drug causes withdrawal symptoms
   (c) Slow it down
   (d) Different penalties for illegal possession; because some are more dangerous than others

Chapter 3

How Science Works

1. contradicted the bible; he knew most people were very religious; worried that they would be very upset
2. some species must have died out; these were likely to be the less well adapted; could find some fossils that were similar to organisms living today
3. a theory is an idea or an explanation; data is factual information; a theory has not been proved correct

Exam practice questions

1. **C**
2. **D**
3. **A**
4. Nucleus; chromosomes; DNA; proteins
5. **(a)** There would only be 23; they would not be in pairs
   (b) The last pair are different; contains an X and a Y chromosome
6. **(a)** Recessive; Jackie and Leroy have the allele but not the disorder
   (b) Leroy's gametes are F and f; correct offspring: FF, Ff, fF, ff
   (c) 25%/1 in 4/a quarter
   (d) Would know for certain if she was expecting a baby with cystic fibrosis; she would then have to decide whether to have a abortion or not

Chapter 4

How Science Works

1. spread over large areas; difficult to track them / find them all
2. two from: only hunt for scientific research; vary the type of whales being hunted; but have caught fin whales that are protected; some people are not convinced that it is for scientific research
3. it would keep their culture alive; they would not kill many whales

Exam practice questions

1. **C**
2. **B**
3. **D**
4. amphibian – moist, permeable skin
   fish – wet scales
   invertebrate – an animal without a backbone
   mammal – covered in hair or fur
   reptile – has dry scales
5. **(a)** Two from: store of food or energy/do not need to eat so often/helps insulate the body from the sun
   (b) Deep roots can absorb water from deep underground; roots that are spread out can absorb water over a wide area; shallow roots can absorb the water before it evaporates
   (c) Large animals cool down more slowly; small ears reduce the surface area
   (d) Less competition between the larvae and adults
6. **(a)** Three from: place quadrat at random/count the number of animals in quadrat/repeat and take an average/multiply the number of organisms by the number of quadrats that would cover the whole field

(b) (i) Three from: ladybirds eat greenfly/ greenfly increased in July because plenty of food and warmth/more food for ladybirds so their numbers increased/greenfly numbers dropped because more were eaten so ladybird numbers started to drop

(ii) predator–prey graph

Chapter 5

How Science Works

'I think that it is wrong...' – Person may be worried that cloned embryos are being destroyed to extract stem cells and may think that this is destroying (potential) human life.
'I am not happy...' – Person may think that it is wrong to only allow this treatment to be available for people who can afford it.
'I think that more people... – An unfertilised egg cannot develop into a person so it is acceptable to destroy it to extract stem cells.

Exam practice questions

1. mitochondria, cell membrane, vacuole, cell wall

2.

	Osmosis	Diffusion	Active transport
Can cause a substance to enter a cell	✓	✓	✓
Needs energy from respiration	✗	✗	✓
Can move a substance against a concentration gradient	✗	✗	✓
Is responsible for oxygen moving into the red blood cells in the lungs	✗	✓	✗

3.

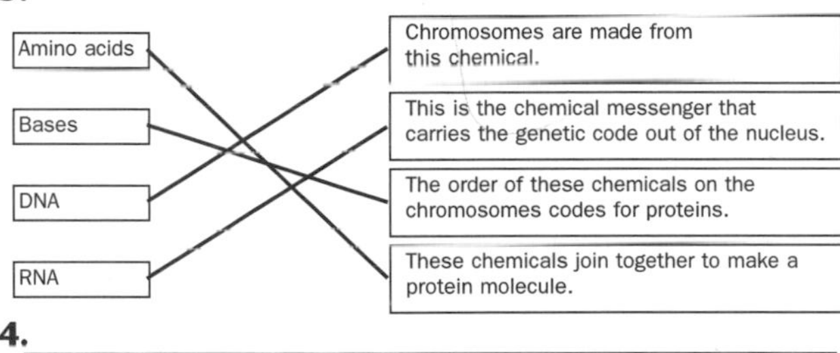

4.

The number of chromosomes in a human body cell.	46
The number of cells made from one cell when it divides by meiosis.	4
The number of chromosomes in a human sperm cell.	23
The number of strands in each DNA molecule.	2

5. (a) On the chromosomes in the nucleus

(b) Each gene has a different order of bases
The bases code for the order of amino acids in the protein

(c) 1094/20 000 x 100 = 5.47%

(d) The liver
Uses the most genes and they code for different enzymes

6. (a) Lives underwater
Makes bubbles of oxygen that can be counted

(b) oxygen

(c) (i) He uses the same piece of pondweed throughout

(ii) He counts the bubbles three times at each light intensity

(d) Use a more accurate timer

Chapter 6

How Science Works

1. More likely to convince people because they believe it is based on scientific fact.

2. The first article tends to concentrate on the environmental effects of farming. It does not consider the health issues.

3. They have different backgrounds such as different jobs. They have picked up different views from their parents. Some may have strong views on animal rights.

Exam practice questions

1. carbon dioxide, magnesium, nitrates, auxins

2.

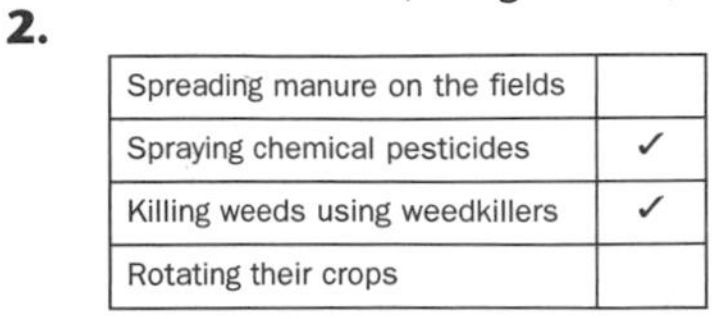

Spreading manure on the fields	
Spraying chemical pesticides	✓
Killing weeds using weedkillers	✓
Rotating their crops	

3.

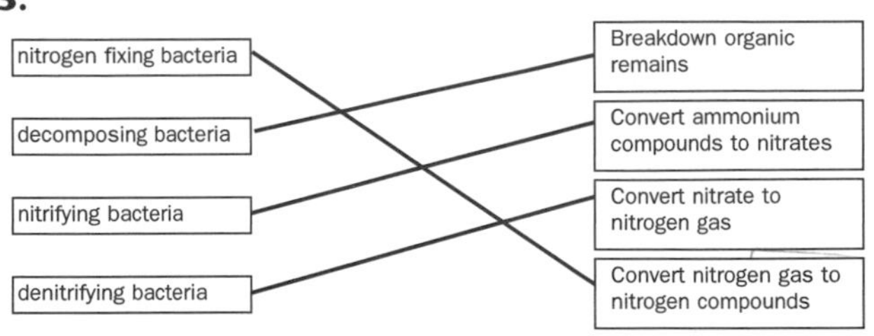

4. (a) 10%

(b) heat / excretion / uneaten parts / egestion

(c) 200 kJ

(d) So much energy is lost at each stage, there is not enough energy left to support another level

(e) Leaves are broken down by decomposers
They give off carbon dioxide, in respiration
The carbon dioxide is used by the trees for photosynthesis

5. (a)

Sales of organic foods are increasing dramatically	F
Organic food is better for people	C
Much organic food is dearer than intensively produced food	F
Organic food tastes better than non-organic food	C
Artificial additives can make food look more colourful.	F

(b) People are worried about the effect on the environment of growing non-organically
They do not want to take the risk that the non-organic food may be harmful

(c) Some countries are struggling to produce enough food
Growing organically may reduce yields / cause starvation

Chapter 7

How Science Works

1. (a) Far more people require a transplant than are receiving them.
The number of people requiring transplants is increasing.

(b) The gap between the number of people requiring a transplant and the number that are receiving them is getting wider.

2. (a-b) The answer depends on a person's ethical views but arguments include ideas about personal choice and also should people be allowed to gain even if they are not prepared to donate.

(c) Possible factors include: the age of the patient, how ill they are, how likely they are to survive, if the damage is self inflicted.

Exam practice questions

1. root hairs, xylem, phloem, stomata

2.

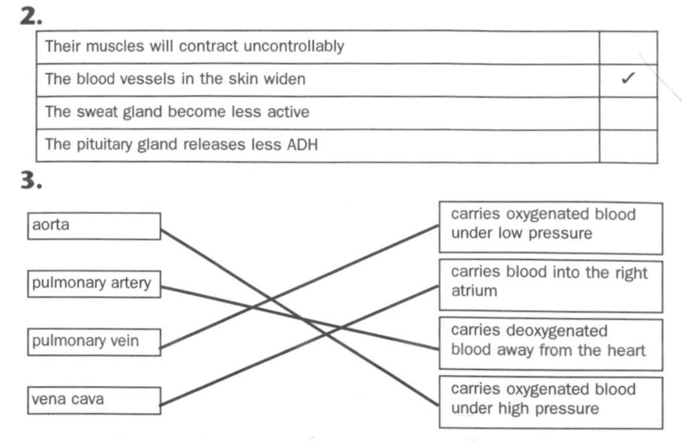

Their muscles will contract uncontrollably	
The blood vessels in the skin widen	✓
The sweat gland become less active	
The pituitary gland releases less ADH	

3.

4. **(a)** digested by amylase enzymes, in the mouth and small intestine, maltose then digested by maltase
(b) **(i)** ten minutes
(ii) Starch needs to be digested.
Glucose needs to be absorbed
(c) **(i)** 280 µg per 100 cm^3
(ii) blood glucose levels rise, detected by the pancreas, insulin released to bring blood glucose levels down
(d) Insulin levels would not rise so high, Glucose levels would rise higher / stay higher for longer

5. **(a)** organs are too complicated / difficulty in powering them / may be rejected
(b)

Some peoples organs could be removed in error	
It would increase the number of organs available for transplants	✓
It would avoid having to ask relatives about donation	✓
Some people have religious objections to transplants	

(c) Forget to do anything about it, say they are willing but really they do not wish to donate

Chapter 8

How Science Works

(a) There is no indication that other factors in the women were taken into account or kept constant.
(b) Asking the women about their eating habits would not give a precise record of their vitamin A intake.

Exam practice questions

1. selective breeding, genetic engineering, gene therapy, genetic fingerprinting

2. **(a)** D, A, B, C
(b)

the process uses hormones to cut DNA	✗
the Bacteria can be grown in large fermenters	
the insulin produced is a hybrid of human and bacterial insulin	✗
the bacteria can be grown on cheap waste products	

3. **(a)**

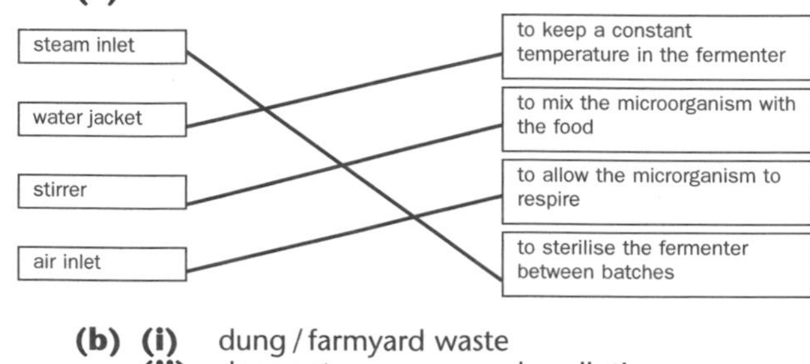

(b) **(i)** dung / farmyard waste
(ii) does not cause so much pollution, sustainable / will not run out, does not have to be transported around

4. **(a)** **(i)** B
(ii) A

(b) No, the method does not take into account how serious each defect is
(c) This is because they have different views on whether it is correct to use genetic engineering
They are trying to back up their argument

Chapter 9

How Science Works

1. **(a)** Diabetes was killing so many people;
People were less concerned about animal rights in that period
(b) To save time because diabetes was affecting so many people

2. Arguments for and against include personal choice; against the cost of production

3. People are worried that the genes will spread to other crops; and worried about possible damage to animals if they eat them

Exam practice questions

1. **1** decomposers, **2** pathogens, **3** hosts, **4** photosynthesis

2.

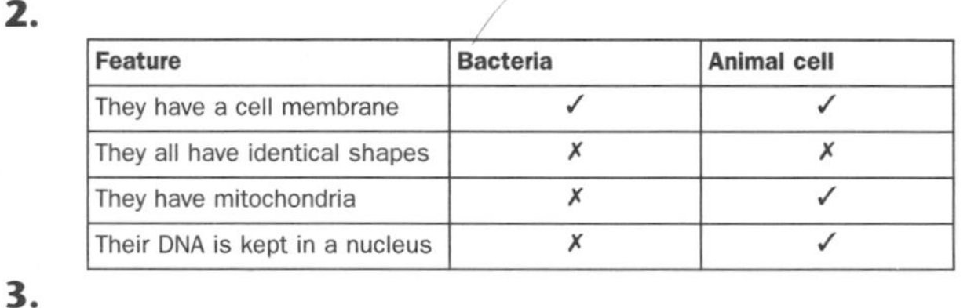

Feature	Bacteria	Animal cell
They have a cell membrane	✓	✓
They all have identical shapes	✗	✗
They have mitochondria	✗	✓
Their DNA is kept in a nucleus	✗	✓

3.

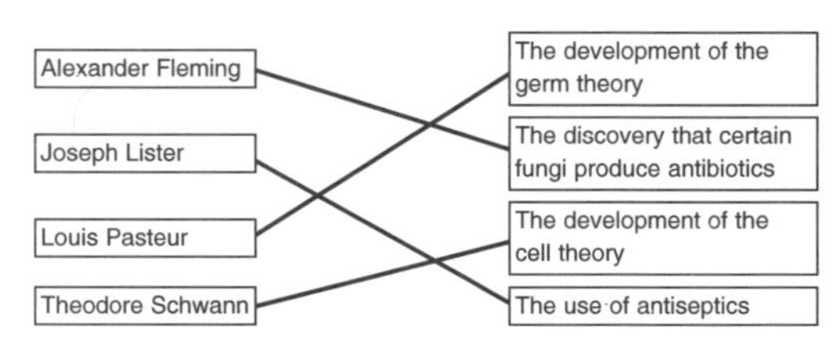

(a) In case of leaks.
To keep it warm.
To contain any explosion.
To eliminate smells.
(b) The bacteria are respiring anaerobically.
(c) It only gives out the carbon dioxide that is taken in by the plants growth.
(d) No mains electricity or gas / difficult to transport fuels.

4. **(a)** **(i)** carbon dioxide.
(ii) To stop air getting in.
Yeast must respire anaerobically / prevent microbes getting in.
(b) **(i)** Keep them at the same temperature.
(ii) With extra sugar the fermentation is faster. But reaches the same concentration of alcohol in the end.
(iii) There is more sugar present so there are more collisions.
The yeast is killed by a certain concentration of sugar so now more alcohol is made.
(iv) The wine with the extra sugar will taste sweeter.

Chapter 10

How Science Works

1. 'I don't like the idea of releasing GM organisms into the wild.'
This person could be worried about all of the treatments;
They all involve GM organisms that could escape into the wild and could harm other organisms;
'I don't think we should rely on one drug alone, this might not always be effective.'

This person is worried about the third treatment;
Populations of pathogens can become resistant to certain drugs
'My worry is that the parasite might mutate and then the parasite that you have injected could be very dangerous.'
This person is worried about the second treatment:
The harmful plasmodium parasite may mutate in a person's body and cause malaria

Exam practice questions

1. **1** osteoporosis, **2** pacemaker, **3** dialysis, **4** amniocentesis.

2.

	Glucose	Protein	Urea
Present in the blood reaching the kidney	✓	✓	✓
Passes out of the blood in the filter unit	✓	—	✓
Reabsorbed back into the blood from the kidney tubule	—	—	—
Usually present in the urine	—	—	✓

3.

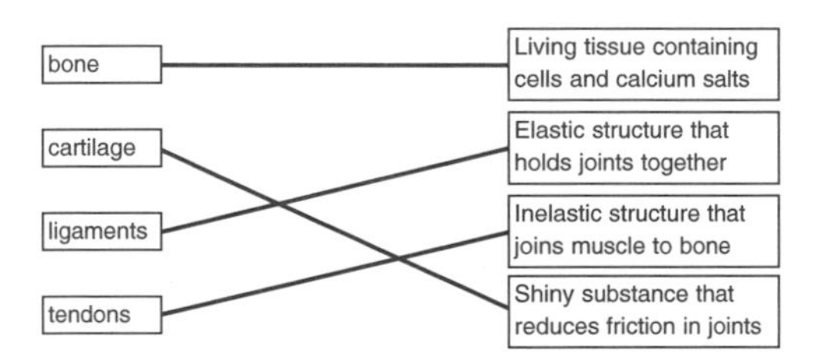

4. **(a)** **(i)** 130mmHg
(ii) 15mmHg
(b) Much thicker muscle wall in left ventricle.
It needs to pump the blood further.
(c) Not so high.
Only has to pump the blood to the lungs rather than all round the body.
(d) Pressure is very low.
Valves stop the blood flowing backwards.

5. **(a)** **(i)** alveoli
(ii) pollen, dust, animal fur.
(b) **(i)** Vital capacity.
Tidal volume.
(ii) smaller tidal volume.
About 1 litre compared to 1.25.
Smaller vital capacity.
About 3.5 litres compared to 4.

6. **(a)** tick in first line.
(b) It tends to occur in poor countries.
They would not be able to afford the treatment.
(c) Reproduces very quickly.
Produces high yield.
Does not take up much space.
Can be grown on waste.

Chapter 11

How Science Works

1. Probably it is a combination of many differences that makes us different to animals

2. They must be careful not to anthropomorphise, i.e. they must not assume that the animal is behaving like a human for the same reason

Exam practice questions

1. **1** body language, **2** pheromones, **3** learning, **4** culture.

2.

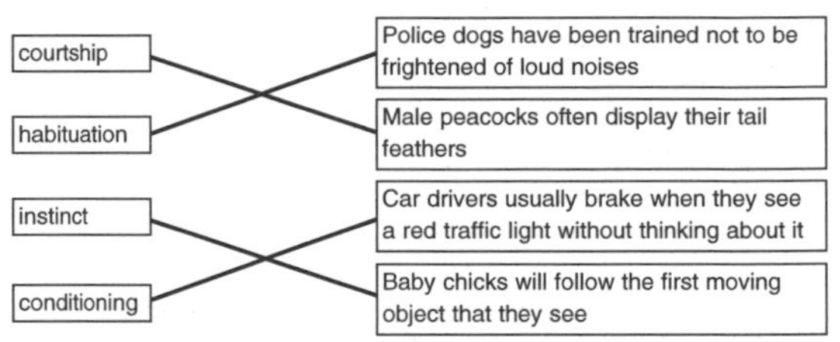

3. **(a)** Goshawks are more successful when there are less pigeons in the flock.
Give an example of figures to back up this argument.
(b) In large groups it is more likely that a pigeon might see the goshawk.
It can then raise an alarm / mass movement might confuse the goshawk.
(c) They are more likely to run out of food in an area.

4 **(a)** It allows a female to choose a male that has good genes.
It indicates readiness to mate.
It helps with bonding.
(b) Ear tufts are bigger or standing up.
Necks are longer.
Bodies facing each other.
(c) To use to build nest.
To form a bond.
To practise feeding young.

5 **(a)** The dog produced saliva.
(b) The dog associated the noise with food.
Conditioned behaviour.
(c) People might think that the experiment was cruel.
Infringe animals rights.
May learn some basic biological facts.
They may become useful in the future for treating disorders.

Index

Notes

Notes

Notes